The Heart of Forestry in America
A Berea College Odyssey

by Clint Patterson,
Berea College Forester

ISBN: 9780578565293 (paperback)
ISBN: 9780578565309 (epub)
All illustrations are by Clint Patterson.

CONTENTS

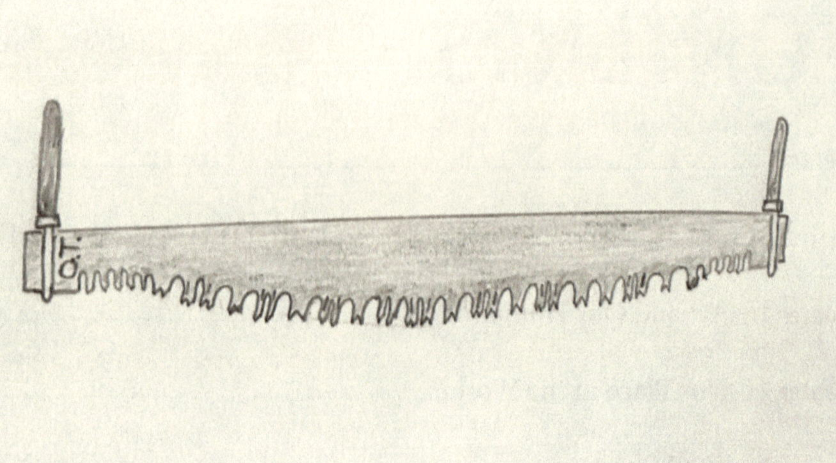

Preface

Brownstown, IL (1990).

The old cherry tree was huge…at least, what was left of it. Hollow as a pumpkin, there was nothing left of it but an immense stump about eight feet tall and three feet in diameter. All but a few small branches had broken out; leaving gaping holes that a child could look through to see the interior of the main trunk. My cousins and I had climbed on these branches, played with toys in the holes, and every spring tried to beat the birds to a few of the cherries that still grew from the sparse twigs.

Grandma used to pick cherries from the old tree before it "went kaput", as she would say. I am pretty sure I ate a piece of the last pie she made from them. The tree stood directly outside the kitchen, and grandma could see it through the many-paned windows above the sink. I remember her standing there, washing dishes, and hollering at birds…more joking than anything…to "get the dickens away from those cherries!"

The old "Arts and Crafts" style house, built in the 1930's, was a sort of landmark in their tiny town; with its multitude of windows, wide overhang, and big front porch. Never modernized, the home was still heated with the original hot water furnace and iron banisters, and had no air conditioning. Hollering at birds through the open window was no longer even called for, and the plastic owl and rubber snake grandma hung it to scare the birds were just decorations now since barely a single cherry had been seen in a couple years from the ever-dwindling live branches. Grandma didn't have a heart to get rid of it until now. She decided it was time to finally get rid of the old hollow gal.

Grandpa only had a two-man crosscut saw to cut down a tree with. He had never bought any of those "newfangled gadgets" like a chainsaw or power tools, and scarcely bought anything at all, actually. Grandma said he as "tight as bark on a tree". It was difficult to not agree, as he came outside in his oil-stained coveralls that you could literally stand up in the corner after he took them off… and his work boots that had duct tape wrapped around them to keep the soles from flopping. I was "back home" from college; visiting them for the week-end…so I got to be the second man he needed to operate the saw.

"Go get the crosscut saw off the wall in the garage", grandpa told me to do, as he finished his black coffee, lit up a cigarette with his WWII zippo and slipped it into the corner of his mouth. I had seen the saw hanging in the garage for years, and knew where it was…but had never used it. As I walked around the house to the basement garage that was under the sun porch, I thought to myself how I used to walk "downtown"…to what used to be a downtown…to pick up the mail and buy a pack of cigarettes for grandpa. Grandma would send a grandkid to do this, and we got to buy some candy for doing it…"just put it on the tab" she would tell us to say. Gene's Red Fox grocery store was now closed, the old Sundries drugstore closed…not much there now. "Here I am, going to college now, and cutting down the old cherry tree. Time sure change." I contemplate.

There it was…..hanging on a nail in the garage where it always had been. Getting it down, I saw the back of it for the first time. "O.T." was painted on one end, near the handle, in red paint. The two inch tall letters piqued my curiosity. By the time I brought the saw around to the old cherry tree, Grandpa was standing there. Wearing his one-piece work overalls with years of crude oil stains on them, he looked like I had always remembered him from my earliest childhood memories. His face was weathered from so many years working outdoors in the wind and sun, but you could still see his handsome features that hinted of Errol Flynn and Clark Gable in their prime. Wiry and tan, prickly grey stubble, oily overalls, worn out work boots, and a "Feller Oilfield Service" cap. Still working in the oilfield into his 70's, he had went from driving a truck in World War II to driving an oil rig in the Loudon Field…one of America's "oil boom" areas…and would keep working out there as long as they would let him. Every time I drove the six miles from St. Elmo to Brownstown to see him, I drove past Feller's Oilfield Service and looked for his truck. Before I could drive, it was located on the opposite side of the road in the old airport building where Charles Lindbergh had often flown to. When Lindy wanted to get away from it all, he liked to hang out in St. Elmo.

Grandpa never seemed to want to get away from the oil field. He knew the maze of lease roads in Loudon Field like the back of his leathery, oil-stained, wrinkled hands. A couple generations of oilfield workers looked up to him like a dad or a grandpa.

"What's the 'O.T.' stand for?" I asked, as I handed the saw to Grandpa. "Otto Torbeck, my dad. It was his saw. Sawed a lot of wood with it. Back then, we all went to the woods in the winter, after the farm work was done. Everyone had a little piece of woods in the Big Woods off away from the German Prairie and we'd go by the wagonload from the prairie to the woods and cut all the wood for lumber and firewood for the winter."

Later, when we went inside, Grandma dug out some photos from the 1920's showing wagon-loads of men going to the "Big Woods"…grinning and posing with their axes and crosscut saws…pointing to each one and telling who they were…mostly uncles and cousins. These photos were such a stark contrast to the stern, unsmiling faces so often featured in photos of that era. Clearly, these folks enjoyed working…especially in the Big Woods.

"Well, we might as well get started, instead of stand around all day", grandpa said as he held out one end of the saw for me to take. Getting in position, he instructed me to pull, but not push, and we began to saw through the tree…whoosh, whoosh, whoosh, whang! Whoops…I accidentally pushed. Whoosh, whoosh, whoosh… It went fast, as there was only a ring of wood to saw through, with the interior all long rotted away. Fine, reddish chips of cherry wood flew from the cut and a pleasant smell filled the air…fresh cherry wood sawdust mixed with burning tobacco. Before the saw started to pinch from the weight of the tree pressing down on the cut, Grandpa drove a wedge into the gap with his ancient sledgehammer. "Thud" went the big hollow stump as we pushed it over. Then, we spent the next hour or so cutting it up into chunks and carrying/rolling them to the corner of the yard where they would be split into firewood.

"Hit where you look, don't look where you hit", he instructed me, when I helped him split the wood. I liked to use the splitting maul, but sometimes missed the spot on the chunk of wood I meant to hit, and Grandpa would shake his head. He preferred to lightly tap in a splitting wedge and then hit the wedge with a sledgehammer… and I never saw him miss hitting that wedge. The wood would be used for the occasional fire in the fireplace; most likely at Christmas time when all of Grandma and Grandpa's kids and ten grandkids would come to open presents and have a big meal.

"So, you learning anything in college?" Grandpa asked as we were cleaning up the old cherry tree remains. I told him a few things…how we learned to measure trees and figure out the board foot volume of individual trees, how to use a prism to do point sampling, to inventory a forest, and how to write forest management plan.

"That's 'how' to do things, but what about 'why' you do them?" he then asked.

"Well, it depends on the goals of the landowner…" I began to reply…the standard response I had learned… as this is what was first considered when we began to do a forest management plan.

Grandpa didn't care for my canned answer. He stopped what he was doing and motioned for me with his wiry, darkly tanned arm to come over to a sugar maple tree in the yard…one of four sugar maples grandma and a couple of their older kids had planted years ago. My cousins and I had grown up climbing in them. "What's the scientific name of this tree species?" he asked me. "Acer saccharum", I replied. "So, what does 'acer' and 'saccharum' mean?" he then asked.

I had never been asked what the scientific…Latin…names actually meant, or even saw any reference to their meanings. I thought he would be impressed that I just knew the scientific names. "Hey, wait a minute, how does he know that?" I wondered. "Well, I would guess 'saccharum' means sugar, because it sounds like 'saccharin', but I don't know what 'acer' means, do you?"

He pulled a leaf off the tree and pointed to the lobes. "The leaves have sharp lobes…points…acer means sharp. Yes, saccharum refers to sugar. Do you know what the wood is good for?"

I replied that it was also called "hard maple" and had a very hard wood, and I knew it was used for bowling alley floors and butcher blocks, among other things where a hard wood was required. Then, I had to ask "So, how do you know all this stuff about trees?"

Ignoring my question, he asked me for an example of a deciduous tree with soft wood that would be good for carving. I replied that basswood was good for that, but not as good as butternut…which was dying off from butternut canker. "What's another name for basswood…you know that?" he quickly asked. "Linden", I answered. "Right…'Unter den Linden' is a famous boulevard in Berlin, named for the linden trees… Hitler had the old ones cut, and replaced with flags…the public had such a fit over it he replanted them… linden means soft, and yes it is a good soft wood for carving."

It seemed incredible to me that the public rose up against the cutting of the linden trees, and got them replaced…amid the backdrop of larger issues that were not sufficiently opposed. I assumed he must have known about Unter den Linden from being in World War II, as I knew he served in the European theater and fought across France, Belgium, and Germany. I had seen the horrible photos he brought back of concentration camps. But, how did he know all this stuff about trees? Again, I asked, "How do you know all this stuff about trees?"

"I went to the parochial school. We all spoke German at home. The school-master taught us in German, then Latin, and English last…from the Latin, so from the Latin we would know the root words. He said if you're going to learn to speak English, you need to first know what the words mean, to know what you're saying…and better than the English-speakers. It's been very useful in life.

"The thing I remember most from school is the schoolmaster drilled it into our heads, over and over, is this: 'order is the highest virtue'. Without order, you can't have anything… otherwise, it's just 'disorderly conduct'. People can get arrested for 'disorderly conduct' you know", he said with a glint in his eye and mischievous smile that almost caused his cigarette to fall out of his mouth. Holding back a laugh, he pretended to cough as he held the cigarette in place.

"As far as learning about forestry, every German is a forester of sorts… or used to be… we all worked and went for walks in the forest, and would rather be doing that than anything else. There would be a Forstmeister in charge… always someone was in charge …organized… a forester should have authority, not be like a servant told what to do by someone who doesn't know…so they decided what trees to cut for what and why and where. It would be stupid to make something like firewood out of an oak that was high quality enough to make a beam for a barn, for example…so we did it all organized across the whole forest area, not "every man for himself" on his own little piece. Everyone helped and everyone learned how. More importantly, though, is to learn why."

Then, grandpa paused for a moment and looked directly at me and asked: "Did you ever read any Goethe?"

He pronounced Goethe correctly, like "Gehr-teh", but I had never heard this name spoken; only saw it in print. He read my look, and went on: "It looks like Go-eth, but is pronounced Gehr-teh. He is the German equivalent to Shakespeare in the English world…no one besides Martin Luther had more influence in Germany. But, he was not just a writer and playwright. He was an artist, a scientist, philosopher and statesman. We had to read Goethe. Goethe taught us why. How can you be a forester without a proper philosophy behind it?" he asked.

I explained that I had heard of Goethe, and his famous work "Faust", but was only vaguely familiar with it…that we did read about Aldo Leopold and his "land ethic" and discussed John Muir and his influence and efforts to preserve forests. "But, what was the result? Nothing, right? Just talk about it and move on, right?" he said. I didn't have much to say. He recommended I read Goethe.

My '79 black Mustang with a hatch back was parked in the yard. Walking around looking at trees, we had walked past it a couple times. I decided to get my things out of the back to bring them in the house. There was an empty red cooler in there with brand name "Gott" written on it. "Gott!? Grandpa exclaimed. "Let me see that cooler", he said, as I began to close the hatch. "Gott means God, that's not right to use that for a brand name…not right", he said; shaking his head. Years later, after Grandpa had passed away, I saw the word Gottesblick in a book about early German forestry and wished I had the chance to ask him if he was familiar with that word and what it meant to him. As I remember it told, before forestry had become academically institutionalized, when a forester was asked how they knew what to do, they would reply: Gottesblick.

I put the cooler back in the car. The next day, Grandpa went back to work in the oilfield, and I went back to college. I had a renewed sense of purpose…to delve deeper into the way of thinking, questioning and reasoning that Grandpa exhibited. I began to read some of Goethe's writings. Here was someone who had been famous for his works in both art and science…and yet, in college, I had had to choose one or the other. My forestry major was housed in the Sciences and I would receive a Bachelor of Science degree…not a Bachelor of Arts degree.

I followed Goethe to Rudolph Steiner and his "spiritual science" and felt that it was no wonder no "land ethic" had taken off since Leopold…as nothing was being taught as a philosophy. It all seemed hopeless. Then I read Wendell Berry's The Unsettling of America. When I realized that here was an author who was still alive and writing… I felt I had found a prophet for today's looming issues and hope for an American "land ethic". Wendell's writings blazed an impression on both my mind and my heart that has not dissipated.

It has now been almost 30 years since this conversation with my grandpa took place. Shortly after graduating college, I ran across a quote by Johann Heinrich Cotta (1763-1844) who is known as a pioneer of modern forestry. In 1816, he wrote: "Three principals exist why forestry is still so backward: first, the longtime which wood needs for its development; second, the great variety of sites on which it grows; thirdly, the fact that the forester who practices much writes but little, and he who writes much practices but little." (Cotta)

I have been working as a professional forester now for about 27 years, and it seems that Heinrich Cotta is still right. Why hasn't Leopold's 'land ethic' blossomed? Why does it seem like 'disorderly conduct' is still, too often, the norm? Since reading Cotta's quote, I have written hundreds of articles in newspapers about forestry and other topics on nature and tried to be a forester who both "practices much" and "writes much". Now that I feel like I have enough experience under my belt to feel like I have earned the "forester who practices much" description, I thought it was time to write a book to share my perspectives.

Along the way…working with private landowners, school groups, college students, and all manner of people…I have also discovered the maxim that "people don't care what you know, unless they know that you care", usually rings true. I have spouted off "facts" to people until I was blue in the face, with barely any results, yet often found that anecdotes and personal stories do resonate with people and sometimes change minds. I have written countless forest management plans that no one implemented, and written countless newspaper articles over the years and tried to be that rare forester who both "practices much" and "writes much"; yet have found that a leisurely stroll in the forest with a landowners often yields more lasting results.

I have come to think of Grandpa's maxim: "Order is the Highest Virtue" as containing a sort of formula which welds together science, art and spirituality. It recognizes an inherent order in all things…an order that is perfect and therefore harmonious. Thus, this order demands both our attention and our affection. Newton and Goethe quarreled over the direction that science would go and even on the very nature of light. Newton sought to understand the natural order of things by reducing them down into their component parts; while Goethe sought understanding through studying how they fit together. If we want to both see the forest and the trees, we should do both.

Forestry is often defined as an "art and a science"; yet I have found that in the field of forestry there is plenty of science being dished out, while the art of it is more difficult to come by. Finding myself in a place largely defined by egalitarian principles, and rich in arts and crafts, I feel that it is time to welcome the arts into forestry, and allow the "spirit of the forest" to infuse us with inspiration to seek new solutions to forest management challenges. This is how I believe we will, as a society, develop the philosophy needed for a "land ethic" to flourish. I can think of no more fitting a place for this to happen than Appalachia: the most diverse temperate forest regions of the world. Within Appalachia, I can think of no more fitting place than Berea College…the first integrated college in the South, one of only a handful of free-tuition work-colleges, Kentucky's "arts and crafts capital", and the birthplace of egalitarian forestry in America.

Introduction

The old Forester's Cabin was moved to this spot around 1910. Sitting at my desk, I can look out the window in front of me and see Bear Mountain…the highest point on the Berea College Forest. An old spring house is up there, along the way to the top. Behind me is our two-year old Forestry Outreach Center…a welcome center to the forest, where educational programs and conference space is offered. To my right, is the parking lot which was once an apple orchard. On sunny weekends, it is packed full of cars; especially since our trail system received the designation of best in Kentucky by Outdoor Magazine last year. To my left, is a grassy pasture with some remnant shortleaf pines…which, along with American chestnut, we are trying to restore on the forest. Both species were mentioned in our original forest management plan, written by Silas Mason as being abundant on our ridges at the time (1905).

The logs of the cabin are painted grey, with white chinking. The dovetail notches are still tightly fitted together. Where the paint has flaked off, the weathered gray wood of oak logs peek through…logs from trees that probably sprouted from acorns dropped before Daniel Boone passed through here. A massive stone fireplace and chimney occupy the living room and the wide, inviting front porch face highway 21 to the south. The highway used to be called Narrow Gap Road…a washed out, rocky path in our earliest photos. Clearing brush from along the old wooden plank fence around the yard, I uncovered some old lilac bushes and wondered if Silas Mason, the original College Forster had planted them. Whoever planted them, now they are free from the vines and brush that had grown up around and over them and almost smothered them to death. Perhaps, next year, they will bloom again.

There is a boot scraper imbedded in the concrete bottom front step of the old porch. I imagine how previous Berea College foresters must have scraped their boots on it and how I will now for a time…until someone else does. A couple of years ago, I interviewed Phillip Harrison and his wife, Fanny here at the cabin. Phillip's father, Howard H. Harrison, was the third Berea College Forester, from 1927-1947. Phillip is 100 years old, yet remembered his childhood here as clear as a bell and had many notable and funny stories to tell. One such story involved his little brother…who would eat all his candy and then steal Phillip's, too, while he was away at school. So, Phillip put a mouse trap in his bag of candy one day and then carefully placed it back in the drawer where it was kept. Well, his plan worked…when he got home, his little brother had a sore finger…but his mom was NOT happy about it!

I used to think of 100 years as a long time. Now that I am just one year shy of halfway there, it does not seem long at all. In the nine years I have been the forester here, I have had the satisfaction of helping us become the first college to enter into a carbon sequestration agreement on their forest, took part in helping our LEED-certified residence hall be honored as the "greenest in the world", and began a prescribed burning program. But, what gives me to most satisfaction is to see our horse-logging program coming to fruition in time for my favorite living author, Wendell Berry, to see this happen.

The Berea College Forest is one of the oldest continuously managed forests in the United States. It may be the first forest established "from the get-go" for multiple use; and it is one of the few managed municipal watersheds in the country. The forest and its first forester deserve to be better known; to have a book written about them. I am thankful to have the support of Berea College's President, Dr. Lyle Roelofs, to practice forestry here in a creative way; both as an art and a science…and to write this book. How do I go about writing a book that will do justice to this place and to its founding forester whose cabin I now occupy? I don't know…but it's about time someone wrote it. According to Albert Einstein:

"Time is what prevents everything from happening at once" and "…the separation between past, present, and future is only an illusion, although a convincing one."

Berea College's forestry program was started at roughly the same time as that of the Biltmore Estate. Carl Schenck's monument to his Biltmore Forest School has the same date on it (1898) as the first year Berea College offered three forest management classes taught by our first forester, Silas Mason. A horticulture professor, Mason had already been teaching a forestry class at the Kansas Agricultural College. He described the class as "perhaps the most advanced piece of forestry teaching done in any American school at that time". It must have been, as Mason was offered in 1895, the position of Assistant Chief under Bernard Fernow, Chief of the Bureau of Forestry (later to become US Forest Service). He stayed at Berea.

Mason remarked that he received the idea of the Berea College Forest while attending the College's annual Mountain Day excursion at the scenic Indian Fort mountain area before it was owned by College. I like to think that Mason chose to use the word "received" when he described how he obtained the idea of the forest because he recognized that this idea came from outside of himself. (Mason: The College Forest) Perhaps it came via his heart.

Part of the Mountain Day tradition was for the entire College assembly of students, faculty and staff to walk from campus the two miles to the forest, and hike up to the top of the mountain before sunrise. At the top, on the east pinnacle, they would listen to the College Choir sing hymns at sunrise. He surely would have attended the choir service prior to receiving his vision. Mason would purchase the first tract of what was to become the Berea College Forest with his own funds…the very location where the choir sings at sunrise. Mountain Day is still celebrated much as it was then, but folks can drive out to the Forest to begin their hike at the bottom of the mountain.

What started out as a vision, rapidly became a reality after he met the remarkable and generous Sarah Fay, of Boston. A lover of forests, she shared Mason's enthusiasm for his vision and became his benefactor. She financed the rapid purchase of over 5,000 acres of land that became the Berea College Forest; thus insuring a perpetual supply of timber, drinking water, recreational opportunity, and forestry instruction for generations to come. Mason's vision has far outlasted his lifetime and become part of the community and its identity: a working forest that is integral to the community and a catalyst for learning, "green" solutions, a local economy, and positive change.

The Berea College community, with their egalitarian philosophy, could not have provided a more worthy place for Mason's vision of a College Forest to come to fruition. It was established in 1855 by Christian anti-slavery activists to offer tuition-free education to both whites and emancipated slaves and was the first integrated college in the South. This egalitarian framework that exemplifies Berea College gave rise to Mason's forestry initiative at a time when forestry practices were mostly the province of the wealthy (e.g. Biltmore Estate) or the government. (A Century of Forestry)

It is an honor to serve as Berea College's tenth forester and to manage this now 9,000 acre working forest. My own vision for this forest is for it to become a place where a land ethic, as described by Aldo Leopold, blossoms and bears fruit from which seeds can be planted throughout Appalachia and beyond. For these seeds to sprout and flourish, they must be fertilized with inspiration and watered with wonder…concepts that lie outside the realm of science, but are ready to be discovered in the arts and received by our hearts.

The idea of writing a book which speaks to both the arts and science of forestry; yet also tells the unique story of Berea College, presented a challenge that would require me to pour myself out completely. I would not only be opening myself up for criticism based on professional knowledge, but also on feelings and beliefs. To really pull this off, I had no idea what would broadly resonate with folks; only what resonated with me over my forty-nine years of life. I reached a point, like Kurt Vonnegut, Jr., where he stated in the preface to his book *Breakfast of Champions* (1973), that as he reached his fiftieth birthday, he felt a need to "clear his head of all the junk in there". About half way through this project, I identified again with Kurt where he said…in the middle of Breakfast of Champions: "Once I understood what was making America such a dangerous, unhappy nation of people who had nothing to do with real life, I resolved to shun story-telling." Thus, I almost abandoned writing my book mid-way. People seem to be polarized almost as much as they were when Berea College was founded. To write a book which seeks to find unity through nature began to feel naïve and a set-up for failure and ridicule. But, then I found unity in my personal life…a new beginning with a new wife. I decided that the "real life" that we are supposed to be living is indeed a life of unity, of harmony, of balance…and maybe my odyssey of finding this will resonate with someone. This is what the book is about and I hope it resonates with you in some way. If it doesn't, then at least you know more about the Berea College Forest.

So, this is the story of the Berea College Forest, but it is my story as well. I will take you down the wandering path of discovery and bewilderment that lead me here…a path inspired by Socrates, Isaiah, Paul, Goethe, Emerson, Thoreau, Leopold, and Berry; and interpreted by Planck, Einstein, Tesla, Feynman and others. I think Aldo Leopold would agree that such is the path that leads to the elusive land ethic he so eloquently spoke about it in *A Sand County Almanac.*

This path may not seem obvious or even logical. I didn't have the time or funds to cut the weeds, briars and tripping roots from it along the way (it has not been professionally edited). Other trip hazards I purposely left in the trail to keep you alert…to knock off your hat or skin your shin at the appropriate time. Like Leopold, I will wander off down the trails of wonder and bewilderment that brought me to my current state of mind and purpose. While his book featured "goose music", mine is more like a "wild goose chase". But, I think this is appropriate; as it requires us to use our imagination. I don't know how to write a book the "right way" but I do have a vivid imagination. And, as Einstein proclaimed:

"Logic takes you from A to Z; but imagination takes you everywhere".

My path is designed to be a Waldkunstpfad (Forest art path)…a popular concept in Germany…where both nature and art can be observed, enjoyed and contemplated. Perhaps the "spirit of the forest" will speak to us along the way. Often, scenes from Grimm's fairy tales are portrayed in these art paths there, and so let us imagine our path is enchanted too. Don't worry if you get lost, just

wander.   Not everyone that wanders is lost, and sometimes we have to wander
to find ourselves. Science is as wondrous as any fairy tale and the art of telling
this tale is as important a role for a forester as any. So, let's explore again with
childlike wonder to see what we find; but, now adults, we will wander purposefully.   If you like the book, please let me know. If you hate it, there is no need
to contact me because I have already advised you in advance to "get lost" and
"take a hike".

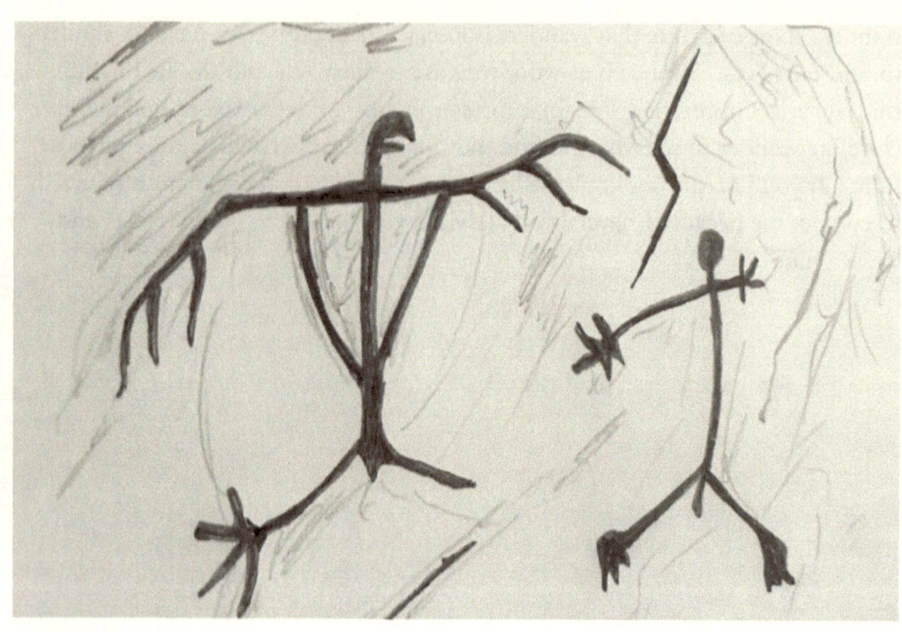

Chapter 1:
Art and Our Human Identity

The artist has a twofold relation to nature; he is at once her master and her slave.
Goethe; Conversations with Johann Peter Eckermann

New Harmony, IN (1982). The sky was a perfect shade of blue... brilliant baby blue with a hint of turquoise. The purest white cotton cloud covered about one-fifth of the roundish shape that framed the sky. I stood there, looking up at the sky and the slow-moving cloud until my neck hurt...imagining the cloud as a living thing as it gracefully manifest itself as a larger portion of the framed enclosure until it moved past and only blue sky filled my eyes. Looking down and around the wooden-shingled inside of the strange chapel, my eyes took a minute to adjust. A bit of dizziness swept across my head as a touch of vertigo caused me to adjust my balance. The warm wooden beams and shingles were a fitting material to frame the sky...and I wondered what the sun would look like overhead through the gaping wood roof of this marvelous open chapel. Could it, too, be contained? Could the eyes even take it in? I knew a tree like this. It was an immense sycamore tree in which myself and three cousins had discovered that was as hollow as Grandma's cherry tree...and broke off at about thirty feet up. We could all climb inside the huge, culvert-like hollow of its trunk and look up to see the round sky peering down at us. This church seemed alive to me, like that sycamore tree. I was inspired to paint that tree and submit it to SAF's Journal of Forestry as a suggested cover. Thankfully, they used it…and I have the framed July/August edition hanging on the wall in my office. The three topics the issue focused on couldn't be more fitting: Aesthetics, Forest carbon, and Restoration Ecology.

Trees are an important symbol in every culture. With their branches reaching into the sky, and their roots deep into the earth, trees dwell in three worlds; heaven, earth, and the underworld: uniting above and below. "Trees of Life" and "Trees of Knowledge" appear in Christian as well as Buddhist, pagan, and other religions. Trees and groves of trees are still regarded as sacred in some cultures.

Since trees are so prominent in Man's spiritual expression, it is no wonder that trees have been a common source of inspiration for artistic expression as well. From ancient Greek and Japanese paintings to Van Gogh's "Avenue of Poplars", trees have formed the setting, if not the focal point, of art throughout the ages.

Trees impart a sense of strength, growth, protection, and grandeur. An ancient, spreading oak tree or a towering redwood, reminds us of how small we are and how brief our tenure on the Earth. These concepts have influenced thinkers and artists throughout time.

Like an artist, a forester is also both master and slave of nature. In effect, he is painting the landscape with his management choices. Choices he or she has to live with. These choices affect not only the aesthetic of the forested landscape in which he works, but the environment at large…its health, its resilience, its ability to sustainably meet the wants of the community…and even the attitudes and inspirations of those who live there.

No wonder children are drawn to climb in trees…

———————

"Get down from that tree and get in the car!" hollered Grandma as she threw her shoes into the floor boards and got in, barefoot. When grandma called me down from "the climbing tree", to go to St. Louis, Missouri, I still had another roll of 24 exposure film for my new 110 camera I had just received for my tenth birthday. I had used the first roll of film when we went to the "crick" (creek)…one of my favorite things to do. I still have the photos of several trees which meant a lot to me. There was the "octopus tree," a large cottonwood with its roots exposed on one side resembling an octopus…the hollow sycamore tree that several of us cousins could fit in a one time. Another favorite, the "kissing trees"; two trees which stood close together and had grown into each other in one spot.

St. Louis was just over an hour away and was where we went to go to a zoo or museums. Grandma's house was like a second home for me. She was always going somewhere and usually took us grandkids along. Grandpa, on the other hand, didn't like to travel. He would say he already had travelled all he wanted in World War II. He would be working in the oilfield or tinkering on something at the house.

Entering St. Louis from the Illinois side…approaching with the Gateway Arch towering ahead, then crossing the Mighty Mississippi…is still an emotional experience for me even though I have done it countless times. The Gateway Arch, built in 1965, and clad in stainless steel, is the world's tallest arch; and the tallest man-made monument in the Western hemisphere. It still strikes me as one of the greatest works of art in architecture…just right for the location and history of the place…in the United States. It still hits me with the same feeling of awe as the first time I saw it…totally unprepared, as I peered out the window as a toddler about nine years after its construction…while standing up in my car seat, as we all did back then.

Stories of Huck Finn and Tom Sawyer, Louis and Clark, and great processions of pioneers entering the West come to mind as the beauty of the place fills the eyes. "I guess I could stand to live in a big city", I thought to myself every time I went to St. Louis, and became tense from the crowds, noise, and traffic, "If I could get away to Forest Park, the zoo, the museums." All these places are free…even now…a rarity in U.S. cities.

Forest Park is a beautiful place for a zoo… and an art museum. The park was redesigned in 1904 for the Louisiana Purchase Exposition; aka the 1904 World's Fair. Citizens were concerned when some 200 acres of the forested park was cleared, swampy areas drained, and the creek rerouted to make room for the Exposition, but the extensive transplanting of trees and monumental, beautiful design of the finished park was widely hailed. Large expanses of native Central Hardwood Forest can still can still be observed in Forest Park today, and some of the oldest and largest native hardwoods in the area can be seen here while exploring the many walking trails.

The park is perfectly designed for the St. Louis Art Museum, as the museum building is in fact the only permanent exhibition facility built for the 1904 World's Fair. Words engraved above the main entrance read "Dedicated to Art and Free to All". A Statue of Saint Louis, King of France, stands outside the main entrance. Originally located at the head of the Plaza during the Fair, it was the first large piece of statuary to greet visitors through the Fair's main entrance. (explorestlouis.com)

Grandma wouldn't let me bring my camera into the Art Museum. Maybe it was against the rules. Who am I kidding…that wouldn't have stopped her…it was probably because she was worried I would take photos of the nude art work. At any rate, I enjoyed the Art museum immensely: discovering first the landscape paintings that I later learned were of the Hudson River School movement. And then there were the Renaissance artists…and the Greek and Roman sculptures. Incredible! I marveled at the skill it must have taken to produce these timeless treasures, and was inspired in so many ways by them to do something great, to learn, to experience! I could have spent the whole day in the art museum, but we zipped through it pretty fast so we could go to the zoo. Even so, it left a major impression on me at this young age, and I think it is vitally important that children be exposed to the arts…and to nature…I was blessed to be able to experience both at Forest Park and this may have set up my concept of thinking from the start that the two…science and art…go together.

———

Die Kunst ist ewig, ihre Formen wandeln sich.
[The art is eternal, their shapes are changing.] – Rudolph Steiner

Ansel Adams (Feb. 20, 1902 – April 22, 1984), the famous photographer and environmentalist, "…did for the national parks something comparable to what Homer's epics did for Odysseus." (Wrote critic Abigail Foerstner in the Chicago Tribune, Dec. 3, 1992). The son of a wealthy timber baron whose fortune was lost in the financial panic of 1907, he passed away on "Earth Day". Perhaps Ansel Adams' transformation and use of photography as an art form, rather than simply a way to record, demonstrated more than anyone else the importance of art as an influence in shaping our attitudes about nature…and, subsequently…in natural resource management and policy.

The first time I saw Ansel Adams' photographs of Yosemite, I was "blown away" by these surreal images. Adams was shy and introverted as a child, and had problems fitting in in school. He ended up being taught at home, and spent most of his time exploring nature in the Golden Gate area of San Francisco where he grew up. At the age of twelve, he taught himself to play the piano, and this was his primary focus and intended profession until he discovered photography…and the Yosemite Sierras.

Adams joined the Sierra Club in 1919 and spent four summers in Yosemite as "keeper" of the club's LeConte Memorial Lodge. Becoming close friends with many of the club's members, who were founder of America's blossoming conservation movement, he remarked that his life was "colored and modulated by the great earth gesture" of the Yosemite Sierra (Adams, Yosemite and the Sierra Nevada, p. xiv).

Environmental activism…in the form of prodigious letter-writing throughout his life…gained Adams the distinction of being one of America's foremost advocates for the preservation of wilderness and a more balanced, restrained use of resources. However, it was his photographs that inspired people and gave him the platform for his advocacy. He wrote to his publisher, Tim Hill, on April 12, 1977, "I know I shall be castigated by a large group of people today, but I was trained to assume that art related to the elusive quality of beauty and that the purpose of art was concerned with the elevation of the spirit (horrible Victorian notion!!)". (http://anseladams.com/ansel-adams-bio/)

The art of Ansel Adams changed not only photography, but attitudes, perspectives, and subsequently…natural resource policy. I like the way Ansel said it…that "art related to the elusive quality of beauty". There it is…beauty… yes. According to Goethe, "Art is constitutive-the artist determines beauty. He does not take it over."

Relating to this beauty, the purpose of art, Adams remarked, is the "elevation of the spirit". Exactly. I would add, that it is my belief that it is when this spirit is elevated, is when it is receptive to the power of love. Then, we can think of others, of a greater good, and endeavors like preserving wilderness and a restrained use of resources become more attainable.

———————

Foresters don't always get to work for landowners or agencies which operate with an "elevated spirit". We often find ourselves caught between opposing forces which push us to generate money from the forest. Sometimes, this is expected to happen without making it look bad. Rarely, do we have sufficient authority in our positions to fully design our own management style or technique to accomplish a "middle ground" unless we own our own land. In the late 19th century, a German forester and forest landowner, Heinrich von Salisch, wrote a book to try to promote just that. He rebelled against the prevailing pressure for economic forestry and its attendant clearcutting, and believed that by simple compromises, a middle ground could be achieved; actually enhancing forest beauty without forgoing income.

Forstästhetik, or *Forest Aesthetics*, written in 1902, elevated von Salisch as the central promoter of aesthetics, trail maintenance, and forest health. The book provides an interesting window into the origins of forestry and landscape design. The book references landscape artists like William Gilpin, Prince von Pueckler-Muskau, and others, to provide a framework for achieving forestry goals while providing a place of beauty as well…seeking to "weld nature and art together imperceptibly". (Salisch, 1902)

Salisch quotes Homer and other poets, as well as sayings he must have grown up with…and is well versed in economics and how forest beauty has a definite value. Whether beauty is adequately evaluated or not, he notes particularly how it affects the tourism industry. Ignoring this value in beauty harms everyone. He also has a sense of humor. One comical illustration features a forester running into a painter and a poet in a forest stand he has inherited management authority over from his predecessor.

The painter and the poet think the forest is beautiful and should not be touched. The forester explains that his predecessor's horrible management decisions have necessitated he clear it all and start over with planted pines. Finally, the poet…admitting he likes what he sees, but knows nothing about it… remarks to the "forestry-minded fault-finder" that "A delusion which makes us happy, is equal to a truth which presses us to the ground".

Salisch observes "The art next to us, the art of gardening, is in the same situation as we are, for even it has to remember that 'the pleasure of the eyes is less charming than an unpleasant sight is disadvantageous'"…. and "After all, we have to accept this limitation, and we have to comfort ourselves that we have some advantages in our field in comparison to the realm of appearance; for example, 'occasionally utility can substitute for the lack of beauty in a real event, but never in a painting".

I have often used gardening as an example to explain forest management concepts to landowners and others. People understand that a garden must be managed to provide food…and it can also look beautiful…particularly if flowers are incorporated into or around the garden. Gardens involve hard work, but many people enjoy this work…gaining exercise, a sense of well-being and self-sufficiency, accomplishment from it. Also, no one opposes harvesting the vegetables or tilling up the garden after it is spent for the season.

The reference to a painting doesn't work so well today. Utility may substitute for beauty in a forest or a garden, but "beauty has no substitute in a painting", according to Salisch…who obviously lived before the modern art era.

In growing as a person, I have grown as a forester. Exposure to art,
literature, and travel to other places is all important to developing inspiration for, and an ethic and a love for, what we do. Foresters, as much as
any professionals, should be concerned with beauty, with harmony, with
function…not only economics. America provided a fertile ground of
wide open spaces and freedom for folks to explore ways in which community could develop in such a way as to bring peace and harmony to
humanity.

Utopian communities have long been associated with both industry
and art…in demonstrating that the two are not mutually exclusive, but
integral to each other. They are fascinating examples of people coming
together with a common vision and purpose to form a better community
and thus a better society. Shaker Village of Pleasant Hill is one such
Quaker community in Kentucky. I have toured their restored buildings
and helped conduct prescribed burns on their extensive lands. This is
meaningful and fitting for me, as exposure to such examples of utopian
experiments was formative to my development.

The second art museum I went to, after St. Louis, was in New Harmony, Indiana. Tiny New Harmony is the site of two attempts to establish Utopian communities. The first attempt, Harmonie (1814-1825),
was founded by a group of separatists from the German Lutheran Church
lead by Johann Georg Rapp. The Harmonists, as they were called, were
quite successful…building a community which achieved great economic
success and was referred to for a time as "the wonder of the west". However, in just over a decade, they sold the town and surrounding lands to
Robert Owen, a Welsh-born industrialist and philosopher who sought to
establish his own communitarian experiment. The Harmonists relocated
to Pennsylvania, and built another community near Pittsburg, named
Economy.

Robert Owen's vision was to create a perfect society through free
education and the abolition of personal wealth and social classes. Now
"New" Harmony, Owen invited scientists and educators to settle in his
Utopian town. While these Utopian experiments both enjoyed success
for a time; both societies, like the Shakers, were celibate and eventually
died out. New Harmony is now a popular tourist destination where art,

architecture, beautiful gardens, history, music and theater, and spirituality can all be explored in one little town with less than 1,000 population, on the banks of the Wabash River. (www.newharmony-in.gov and www.visitnewharmony.com)

Just south of New Harmony, hugging the banks of the Wabash River, lies Harmonie State Park, a great place to explore nature while in the area. I was a District Forester for the Illinois Department of Natural Resources for nine years and my district covered six counties on the other side of the river; beginning with Wabash County, a few miles to the north. This part of the country exhibits some of the highest quality forest trees anywhere. Beall Woods State Park and Nature Preserve, on the Illinois side of the Wabash River just north of New Harmony, in my former District, is one of the few remaining tracts of virgin timber east of the Mississippi River and contains several Illinois state champion trees. I was fortunate to discover a persimmon, *Diospyros virginiana*, tree while working on private property near Beall Woods that turned out to be the National Champion persimmon tree.

The one piece of modern architecture and art that did inspire me at New Harmony was The Roofless Church. Designed by renowned architect Philip Johnson, the non-denominational church is open to the public and is operated by the Episcopal Diocese of Indianapolis. This unique architectural monument was commissioned by Jane Blaffer Owen, founder of the Robert Lee Blaffer Foundation. A resident of New Harmony, she was passionate about preserving nature. She and Johnson's vision was that of a church where the only roof large enough to encompass a world of worshippers is the sky. Johnson was a major figure in the glass and steel skyscrapers of the 20th century in America, and some of his renowned glass and steel creations includes the immense Chrystal Cathedral in Garden Grove, CA.

The appearance of the unique little church in New Harmony is that of a dome shape covered with wooden shingles. The dome is so uniquely shaped… with folds that seem to give the appearance of draped fabric… and at the top of it is an oculus, or round opening. Directly under the opening sits a statue by Jacques Lipchitz, the famous Cubist refugee artist. It was also commissioned by Jane Blaffer Owen.

The Lipchitz sculpture, a bronze of the Virgin Mary, is titled "The Descent of the Holy Spirit". On the back of the sculpture is an inscription, in French, which translates: "Jacob Lipchitz, Jew, faithful to the faith of his ancestors, has made this virgin for the good will of all mankind that the spirit might prevail." Two other originals of the statue exist; one in the Roman Catholic Church of Assy, Haute Savoire, France; the other in the Ecumenical Abbey of Iona, Scotland. (www.visitnewharmony.com)

I found another quote by Lipchitz that I thought went along with his inscription: "Copy nature and you infringe on the work of our Lord. Interpret nature and you are an artist." (The artstory.org)

Lipschitz's "The Descent of the Holy Spirit" sculpture of Mary looked nothing like the more realistic paintings and sculptures of old...many also of the Virgin Mary... I had seen in the St. Louis Art Museum. I didn't understand, at the time, what Cubism was or why it didn't look as "real" as the old sculptures, but I liked it. I didn't remember much of what I read when I first visited New Harmony. I was around twelve and didn't think about the ideas behind the design of the Roofless Church as I gazed through the circular hole in the roof at the clear blue sky that day. I didn't have to. My five senses took it all in...the slight breeze, the almost imperceptible sound of the breeze flowing over the shingled aperture of the roof, the scent of flowers and wood, and the slow-motion dance of light and shadow.

"Interpret nature"...hmmm. It seemed like an instruction given to me. "What does it mean to 'interpret nature'?"

I didn't like the modern art at either museum. I saw them as intrusions...even poison... to an otherwise wonderful place. I didn't know how to describe the ill feeling the modern art gave me at the time, but later when Grandpa told me his maxim "Order is the Highest Virtue", they came to mind as representing disorder. The ideas presented by the roofless church piqued my imagination and prepared me for what was next.

"What are you going to be when you grow up?" Every kid is asked this, repeatedly, I suppose. "A truck driver", I thought...thinking of my Grandpa Torbeck's oil rig that I would eventually learn to drive a stick shift with. I had already ruled out being the next Tarzan by then, but hadn't thought yet about being a veterinarian. Then, I wanted to be an eye doctor; hoping there was something that could be done about this accursed myopia I had from an early age. I was convinced there had to be a cure, as hardy anyone in previous generations in my family was near-sighted. But, it seemed, the thing I was best at was drawing. Maybe I can make a living doing that. But, what I really enjoyed was being in the forest…nature…not sitting at a desk. What jobs are there where you can make a living traipsing around in nature?

My discovery of forestry as a profession came about from a devastating timber theft on my other Grandpa's property. Grandpa Patterson's woods was out in the oilfield north of town. About the time when I was having to fill out applications for college, we went there together to go ginseng hunting. Ginseng *(Panax quinquefolius)* is an herbaceous perennial plant native to eastern North America and used in Chinese medicine. The use and cultivation of Asian ginseng *(Panax ginseng)* dates back to ancient times; perhaps 5,000 years ago. It is revered for its supposed strength-giving, rejuvenating affect, and the human-like shape of some plants' roots became a powerful symbol of divine harmony on Earth. The English name for this herb was derived from the Chinese renshen (Rén-meaning man and Shén- meaning a type of herb). The botanical name Panax was given to it by Linnaeus, the "father of botany", and was derived from the Green word meaning "All Healing".

By the 3rd Century, China had to import ginseng from Korea to meet demand, and by 1716 the North American version of the plant began to be commercially harvested after the French Jesuit priest, Father Lafitau, who was living among the Iroquois Indians, found that a similar plant to the Chinese herb was found in North America. (www.ginsenggeek.org)

Poaching and overharvesting of ginseng has since greatly depleted its populations and threatens its survival. When Grandpa dug ginseng, or "'sang", he only dug the large, mature roots…a plant had to have a least three "prongs" (compound leaf cluster). It was a thrill was to find a "four-pronger" or larger. He only dug them as was legal, and made sense…in the late summer/early fall after their seeds were ripe. He left all the smaller plants to grow and would plant most of the seeds nearby… but carry a few in his pocket to plant in other spots to help them spread. On several occasions were hunting ginseng, Grandpa would stop what he was doing and just look up into the tree canopy and stand in silence… listen…breathe deeply, and comment: "This is where I feel close to God…this is church to me. Do you feel that?" I did. "Yes", I would remark, and point out something particularly inspiring. He would sometimes then begin to talk about Native American beliefs on Nature.

"They believed in a Great Spirit", he would remark and begin traipsing through the woods again. I remembered the name of the sculpture in the Roofless Church…called "The Descent of the Holy Spirit"…and imagined Him descending on us as we stood there. It also reminded me of a Little House on the Prairie episode I saw that resonated with me. It showed a Native American praying before he shot a deer with his bow and arrow. This was juxtaposed with Charles Ingles leading his family in prayer before eating deer. We saw commonalities between their beliefs and what he read in the Bible…commonalities I rarely heard others talk about. It brought comfort to me to know he thought this way…and it gave me strength to speak on nature in spiritual terms.

Sadly, it seems that fewer ginseng hunters follow a conservation-minded approach and will break the law and hunt it in the middle of summer before the fruits are ripe…to dig it before someone else finds it. Often, they dig every single plant…even the tiniest inkling of a plant… wiping out its presence from many a forest. Here is another case where a "land ethic" is needed.

On the way to woods, in Grandpa's 1970's faded red Ford F-150 pickup truck, we followed the route we sometimes took when riding his horses over there…Appaloosas: lanky, speckled-butt horses with thin manes and tails and streaked hooves. The Nez Perce Indians had developed this hearty breed. I think this was why Grandpa preferred them. I wondered if he had loved Native Americans before he loved Grandma or if his love for her had lead him in that direction. She didn't know what tribe her ancestry was, but no one would second-guess from her appearance that she was Native. Her uncles; especially, Grandpa remarked "looked like they walked right of the reservation". The Patterson patriarch had come from Scotland. Some of his ancestors settled in Botetourt County, VA, fought in the Revolutionary War, and were among the first to travel through the Cumberland Gap…settling near Keavy, KY and Covington, KY before moving on to Alamo, IN and then Shelby and Fayette Counties in Illinois. The most noteworthy of them was George Fruits…who fought Indians with Daniel Boone and by some accounts was the last surviving Revolutionary War veteran.

"Did I really want to go to college at all?" I asked myself, as I imagined how there were so many things I could learn from each of my Grandpas and Grandmas and other "old timers" worked out in the oilfield and on farms around here? I didn't know how to do hardly anything, I realized, as I compared myself to the "Farm Boys" and "Gear Heads" that went to my tiny school. Some of them had been driving tractors since they were ten. They could build a house; fix up an old car; overhaul an engine. Living in town, and with relatives who worked instead of farmed for a living, my exposure to learning how to do things had been so limited. School then did offer some auto shop and farm-related classes, but if you took the college-prep route, you couldn't take the shop classes. I think is unfortunate, as it creates a situation where those who go to college are much less likely to know how to actually do anything…yet are likely to be the ones telling those who do know how to do things, what to do!

"Well, my family doesn't have a bunch of land for me to farm anyway, and the oilfield is dwindling out, so I guess I better go to college…sure don't want to work in a factory", I thought to myself…still wondering what I would go to college FOR.

I watched Grandpa Patterson with admiration as he steered the old truck around deep, mud-filled pot holes that had dual-wheel tracks through them…probably from my Grandpa Torbeck's oil rig truck. His muscular arm rested on the top of the window of his truck; his tough hand loosely gripping the mirror bracket. Grandpa Patterson would turn his head and spit tobacco out the window from time to time. Redman. His faded Caterpillar cap denoted where he had worked for years. He never talked about his work…hated working in a factory…had to get out to the woods to unravel from it. He was a horseman…always had Appaloosas and knew how to break them, ride them, and care for them. A mix between Clint Eastwood and Robert Conrad in appearance, he wore cowboy boots and jeans and usually had about the same amount of beard stubble as my other Grandpa. He and Grandma lived in the one-room schoolhouse he had went to school in…across the road from where he grew up. They raised five children there. One bedroom downstairs and a loft above for the kids. He knew these roads and the woods around here as well as anyone. I loved coming to the woods with him…

My pleasant thoughts came to a screeching halt as we dodged a large pot hole in the gravel lease road and crossed his property line. I could hardly believe my eyes. Grandpa's normally calm, stately demeanor changed immediately to an adrenaline-filled anger as the scene unfolded. He stopped the truck and swung open the door in one motion; then took a few steps into the destruction before us. I jumped out the other door and stood there trying to deny the reality my eyes were sending to my brain. Where there had once been mature sugar maple, walnut, oak and hickory trees there were now freshly cut stumps and twisted, bent over saplings. Intense sunlight flooded the whole landscape before us. It was instantly obvious that the whole woods was wrecked…not just this first portion we came upon.

My stomach leapt into my throat as I fought back the impulse to vomit. I looked back and forth at the destruction and Grandpa; waiting to see what he would do. I could see him fighting back tears at the same time that his complexion grew red with anger as he alternated looking up…glaring through an Eastwood-like squint at the destruction…then looking down and clenching his fists. Instead of finding ginseng that day, we found stumps and tree tops. "It's a wonder no one has shot that _______" Grandpa said, in a quivering voice. As we walked through the tangled mess of mangled tops and across deep ruts to survey the damaged, he explained how he knew who it had to be who stole his timber. Not long ago, he had been approached by a man to sell his trees. He told them no. But, the guy was a notorious timber thief who had been known for years to steal trees. The property was remote out here…no houses nearby, and on a bad road. Grandpa worked all week in Decatur…too far away to come out to the woods after a work day.

The only trees left were worthless ones…the woods was "high grad-ed". The practice of "high grading" is far too commonplace, and essentially involves the cutting down of all the best trees and leaving the rest. The opposite of the "worst first" approach which gradually improves the forest, high-grading has the opposite effect. "Can't you do anything about it!?" I gasped, as we crossed the 50 acres and gasped at the mess left behind where there had been…despite the labyrinth of lease roads and power lines cut through the area…and salt kills along the creek from well leaks…a mature and attractive stand of oak, maple, black walnut.

"Oh, I could call the State or something, try to get a value on the timber, but nothing sticks, can't prove anything. They just about always get away with it" he muttered, bitterly. I looked into it a bit, later, by walking into the local USDA Farm Service building and asking around. I was given a pamphlet found with information about the Illinois Division of Forestry on it. This is how I found out that there are foresters who work for the state…called District Foresters in Illinois…whose job is to assist landowners with managing their forests. Programs like this exist in most states, are a good place to start if you are interested in managing your forest. A forester will start out by helping you to get a management plan, and can direct you for further assistance…including, sometimes, cost-share programs may help qualified landowners with expenses associated with managing their forest. Unfortunately, many states…especially Illinois…have not maintained adequate spending for these programs, and have lost personnel pretty much across the board.

"That's what I could be. A forester." I thought, as I walked out of the Farm Service building.

———————

Carbondale, IL (1988). Mom told me to be sure I was back from the Art Department in time for the drawing. "You have to be present to win!" she said, as I rushed off. My application that had Forestry filled in as my chosen major had been received and processed before the one I mailed the same day that declared Art as my major. It was a hot, muggy day at Southern Illinois University and by the time I made it across campus to the Art Department, I was sweaty and out of breath. I walked down a long hallway looking for a reception area or someone to talk to. The first open door I came to, I walked in and was shocked to see the floor was covered with broken glass! I came out of the room and remembered that as I had entered the building, there were a lot of barefoot "artsy-looking" students sitting on the floor. Concerned that someone might walk into the room barefoot, and cut their feet on the broken glass, I went to alert them. As I approached the students, none of them looked up at me or listened at all to what I was saying. A couple of them smirked. I had never been treated this way and began to get angry and feel bewildered.

"Can I help you with something?" I turned around to see an older man…also barefoot…who had quietly walked up behind me as I had been warning the students of the glass. After I explained to him my concern for the broken glass, he informed me… in a staggeringly contemptuous tone of voice… that what I was describing was actually an "art exhibit". Several students then snickered in unison and I felt my face turn hot and red as I fought back the urge to tell them all off. Remembering that I had to get back soon for the drawing, I just took off at a fast jog toward the Forestry Department and never looked back.

My mom gave me a nervous and irritated look as I got back…just in time for the drawing. My name had been put in a hat; along with other students (presumably more than two!) who had an ACT score of 25 or higher. Someone else's name was drawn as I was catching my breath from the brisk jog back. I calmed my breathing down, and listened intently for the next name to be called. If my name was called, I could receive a scholarship for free tuition that semester. The scholarship would be renewed for subsequent semesters for up to two years; as long as our grade point average remained over a 3.0. The scholarship from the hat drawing was for forestry majors only. The fact I was even considered for the drawing was fateful; as I had mailed in two applications for registration at the same time: one listing art and the other listing forestry as my major.

My name was read off and I knew right then that I would be a forestry major. After receiving the scholarship from the hat drawing, there was no changing it now. It is highly unlikely I would have even followed through with going to college if my name hadn't been drawn. Neither my mom nor my step dad had been to college. I knew they hoped I would graduate college, and they would make some sacrifices to enable that to happen. But, I didn't want to burden them. Also, I was tired of school and eager to learn to DO things; to experience some freedom as an adult. Living in a residence hall, under supervision, didn't feel like being free. I wasn't even allowed to have my car at college.

"No way do I belong in the Art department", I thought, after my brief but shockingly negative experience touring the art department. "But, do I belong in Forestry?" I wondered, as I followed through with entering college. After a couple years, I was still drawn to the idea of being an artist of some sort. Foresters use tree marking paint to paint the trees with to designate which ones to harvest. I envisioned this action… managing a forest through deciding what is to be done with it via marking the trees with paint, as creating a work of art. But, there was little room for creativity in forestry, it seemed…even though the textbook in my introduction to forestry class defined forestry as an "art and science".

How could I "paint" anything beautiful with tree marking paint though? A timber harvest usually isn't beautiful…typically, the forest is beautiful before the harvest, and ugly afterwards. I was told forests need "disturbance"…which culminates with clearcutting to regenerate the stand. But, this seemed quite disturbing to me! Art itself was disturbing to me and ugly too…it seemed like whichever profession I chose, I would be expected to turn out something ugly, not beautiful. Why was modern art ugly?

____

The book that later helped me interpret what I was feeling, since I was ten years old and saw modern art for the first time in a museum, was *How Should We Then Live* (1976). Written by Francis Schaeffer (1912-1984), American Evangelical Christian theologian, philosopher, and Presbyterian pastor, Schaeffer was (fittingly, for me) pastor at Bible Presbyterian Church in St. Louis, MO (1940) where he and Edith founded "Children for Christ", after bringing the first "vacation Bible school" program to St. Louis. They went to post-war Europe to minister, and eventually founded L'Abri Fellowship in Switzerland in 1955.

His book *How Should We Then Live?* provides an analysis of Western thought from Roman times in three areas: philosophic, scientific, and religious. The book analyzes how art and architecture reflect changing patterns of thought. The central premise is that a society based on an infinite-personal God provides an absolute that leads to "freedom without chaos". This idea of "freedom without chaos" seemed like the same concept by grandpa had tried to articulate in his "order is the highest virtue" maxim, versus "disorderly conduct" without it. In contrast, according to Schaeffer, when we base a society not on some "higher order", but on humanism, it leads us to despair and alienation. This fragmentation is expressed in the arts in works such as *Les Demoiselles d'Abignon* (originally titled *The Brothel of Avignon*) of five nude prostitutes in Barcelona, by Pablo Picasso (considered to mark the beginning of Modern Art...is proto-Cubist, seminal in development of Cubism and Modern art.

Cubism. Yes. Jacques Lipchitz, the famous Cubist Jewish artist who created the sculpture in The Roofless Church in New Harmony Indiana, was a member of this group of artists...along with Picasso...who forged the Cubist and Modern art movement.

According to Schaeffer, modern relative values are based on Personal Peace and Affluence. He warned that when we live by these values, we will be tempted to sacrifice our freedoms in exchange for an authoritarian government to provide these relative values. This government, he predicted, would not be obvious like the fascist regimes of the 20th century, but based on manipulation and subtle forms of information control, psychology, and genetics.

"There is a flow of history and culture. This flow is rooted and has its wellspring in the thoughts of people. People are unique in the inner life of the mind – what they are in their thought-world determines how they act. This is true of their value systems and it is true of their creativity. It is true of their corporate actions such as political decisions, and it is true of their personal lives. The results of their thought-world flow through their fingers or from their tongues into the external world. This is true of Michelangelo's chisel, and it is true of a dictator's sword." -Francis Schaeffer, *How Should We Then Live?*, Ch. 1.

Picasso, so central to modern art, seemed to fit Schaeffer's theory. He joined the French Communist Party in 1944 and in 1950 received the Stalin Peace Prize (later renamed the Lenin Peace Prize) from the Soviet government. (Eakin, 2000) Schaeffer said "And with truth comes beauty and with this beauty a freedom before God." (Art and the Bible). I spent two weeks in Ukraine in 2017 and saw the many massive collectivist housing complexes built there for people to live in…under a Godless system…all equally as ugly as the next…in a land with just as many exquisitely beautiful Eastern Orthodox churches built long before the ugly housing complexes.

Picasso is quoted as saying "Art is the lie that enables us to realize the truth". So, the same movement that propelled Picasso to fame is defined by his painting of five ugly whores… yet Lipchitz chose instead to sculpt a cubist version of the Virgin Mary. Perhaps Salisch's poet had a point: "A delusion which makes us happy, is equal to a truth which presses us to the ground". This belies the question: "what do we do if we find ourselves pressed to the ground?" Get up? Pray? Or just stay down there, mired in the mud? Like the lotus flower which rises up as a thing of beauty from the mud, let us rise up and blossom… and even the mud can provide a medium for our art.

Art is defined mysteriously, not precisely, and its overriding theme is one of creation. What makes us human, if not the capacity to create? Paleolithic cave men and women were artists…using mud, ash, and other materials to create works of art on cave walls. The magnificent cave paintings of Lascaux Cave, in southwestern France, attest to the artistry inherent in the human race. Near the village of Montignac in the Dordogne region, around 600 paintings of mostly animals occupy the labyrinth of walls and cathedral-like halls of this ancient cave system.

Dated to c. 17,000 – c. 15,000 BC, the paintings were clearly created by skilled individuals. If it looked like "modern art" how would we know humans with intelligence had done it? Both painted and engraved into the uneven walls, the artists worked with the edges and curves of the cave walls to enhance their work. Horses are the dominant composition, with deer, aurochs, ibex and bison, as well as lions bears, and at least one human also present.

Lascaux Cave was discovered in 1940 by four boys who looked into a fox hole their dog had fallen into. Along with some other nearby prehistoric sites, Lascaux was added to the UNESCO World Heritage Sites list in 1979. (Lascaux Cave, by Emma Groeneveld, published on 06 September 2016, www.ancient.eu) From even a cursory view of the Lascaux Cave paintings, the artistry of humans and their natural concept of beauty are manifest. An interesting concept to explore, in this regard, is the perspective from which various nature paintings are created. Is the artist part of nature, for instance, or separate from it?

Classical Asian art often features stylized landscape scenes; sometimes including pagodas and people in them. Once can hardly miss the importance of trees, mountains, horses, fish and other natural elements in Asian culture when exposed to their art. Traditional Western landscape painters mostly painted scenes of nature from afar; as if standing on a high lookout point. During the late 1800's, rural painters began to enter into forested areas and paint the trees and undergrowth from a near-present perspective. This woodland scene genre, called "sous-bois" (meaning understory in French) became prominent as artists from the Barbizon School and Impressionists focused their attention on it. Forest settings of this style are the subject of paintings that Vincent van Gogh made from 1887 through 1890.

Surely, Rudolph Steiner is correct. "The art is eternal, their shapes are changing." Changing with time and place, but eternal. In our own time and place, what shapes will we make? Do we see ourselves as separate from Nature; viewing it from afar? Or, do we see ourselves as an integral part of Nature; vitally connected to it and experiencing it first-hand?

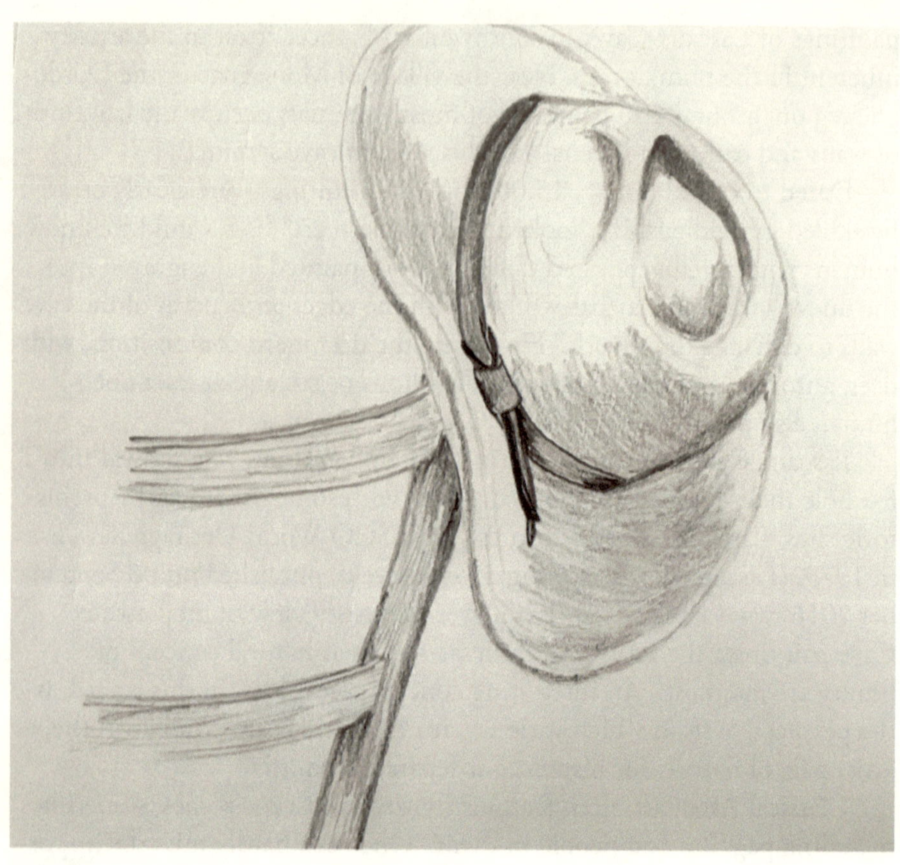

Chapter 2:
Our Place in the World

"Deep roots are untouched by frost." – J.R.R. Tolkien

The old barn still stands. In its loft is where I saw my first baby vulture…a cute little ball of white downy fluff…with a hideous, hissing face on one side. It scared the crap out of me as I climbed the loft ladder and there it was about eight inches from my face…hissing at me. Old, hand-scythed hay likely cut in the 1920's by one of Grandma's relatives, still lies scattered across the dusty loft floor. As my eyes acclimate to light in the loft…I see, through the slivers of sunlight coming through the cracks between the weathered boards, the tracks of the mother vulture in the dust. Then, scanning he gaping cave of the loft, I see and identify raccoon, opossum and mouse tracks from multiple decades crisscrossing the floor in varying depths according to their age. Dry as a bone, still, the roof is intact and the hewn and pegged beams as sound as ever.

The barn no longer is owned by anyone in the family. The trees for the barn must have been cut and hauled here from the Big Woods and built by some of the men whose smiles I had seen in Granma's old photographs. Her mother, Grandma Meseke, was a vague blur in my memory, along with her Uncle Charlie who lived on the other side of Flat Creek. Huge white ash trees…now dying from the invasive pest emerald ash borer, stand between the barn and the creek. I wonder how big they were when Great Grandma was my age. Only a few years ago, she still pole-vaulted across the creek to visit her relatives who lived on the other side of the creek. Grandma showed me the stick, still leaning on the tree where she had left it. The old house had just been torn down. After she passed away, Grandma kept her mom's old wooden shoes on a shelf in the living room where all of us grandkids tried them on at one time or another and clonked around the house in them. I outgrew them when I was about ten, but somehow ended up with them. They sit on my book case in my living room.

Granma would "sneak us in" to such places…no longer owned by family members…so we would know them. We even had to sneak into the cemetery, as the farmers had plowed over the access road to it and right up to the grave-stones. The new owners of such places bought them based on how many rows of corn and soybeans they could plant, how much government subsidy they could get, or because of some tax benefit…not because they loved this land. Fewer and fewer acres are owned by folks with a connection to the land, and a love for it, it seems. A couple years ago, I saw the farm advertised for sale online, and have been afraid to go back and look for the barn since it sold again. It was a miracle the barn had not been torn down by now. The barn is/was the place I have, over the years, imagined myself going when I felt overwhelmed by the cares of this world, of life. To know the barn is not there would wring out my heart like an oil rag.

One of my favorite writers, Wendell Berry, often invokes what he calls a "sense of place". The idea is that human experience, tradition, and history can and do to make certain locations special to those who are connected to that experience, tradition and/or history. People who feel a strong sense of place are emotionally connected to a certain piece of land, house, town, etc.

Before the Industrial Revolution, when most people lived an agrarian lifestyle, it is easy to imagine that most people had some "tie to the land" and a strong sense of place. Immigration and emigration, industrialization, war, and many other factors might have sometimes disrupted this tie to the land for some individuals, but at a pace slow enough for new ties to develop.

In today's world, with its high tech communication, fast pace of life, and mobile workforce, it is increasingly evident that the whole concept of being "tied to the land" or having a strong "sense of place" is becoming antiquated and irrelevant for many, if not most, people. Folks move away from their home town to get a job, raise their children in consolidated, standardized schools until they move away to far off colleges and then move again for employment and likely marry someone from another location.

The quaint idea of staying on the family farm, or even having a "home town" and planting "roots" somewhere may not be what everyone wants… but for those who do, it's increasingly difficult to maintain on a generational level. Until I left Illinois, this is the kind of life I wanted, and wanted to make possible for my children. But, without inheriting a large, profitable farm, that is pretty difficult to do. I tried buying land several times, and never could "swing it", and had to sell out. Finally, I decided to "go where the job is" and realized my children would likely do the same.

Through moving around, I have come to realize that wherever a person ends up… whether they stay on the family farm or move to another country… the ability to develop a "sense of place", is still important. The logic runs both ways. What it all boils down to is that if people matter, then the place they live matters too. The land and the local economy which supports them matters. The environment matters.

Technology, too, could be a positive influence on helping us to develop a broadened "sense of place" and thus a responsibility towards not only where we live, but for our planet as a whole. Much as the astronauts were struck with awe as they viewed the Earth from space, Google Earth now shows us this same reality. Hopefully, as today's mobile, high-tech lifestyle continues to shrink the perceptual size of our planet, a sense of place will not be lost, but broadened so that everywhere will matter. I suppose, then, everywhere would be made to be a nice place to live.

This place called Appalachia is like no other region of the United States. Ever since I discovered the Foxfire books, around age twelve, I have been fascinated by, and interested in, Appalachia. I wanted to learn how to play the fiddle, make an oak basket, smoke a ham, and whittle a wooden chain. I wanted to somehow "become Appalachian". When I would look at a road atlas and daydream about where I would like to live, West Virginia usually emerged as a leading spot… because it was in the center of Appalachia; home of the most diverse and lush temperate forestland in the world. And, it was relatively unpopulated, and I thought maybe I could afford to buy a large tract of forestland there.

I daydreamed about the culture and landscape of Appalachia for a long time before I knew anything about its problems. First, coal mining and its effects on the environment and the people got my attention. Then, I heard about church groups going to Appalachia to deliver clothes, and realized that people must be pretty poor in Appalachia; maybe it shouldn't be on my short lists of places to move to.

When the offer came from Berea College, right on the edge of Appalachia, and I became aware of its commitment to preserve and protect Appalachia and what it stands for, my daydream came back and could become real. I was being offered my "dream job" to become forester for Berea College and move here. So, I suppose I re-discovered Appalachia. Yes, Appalachia does have problems, but that is no reason to stay away… to the contrary, it gives an even greater reason to be here, to be a part of the solution. Here at the edge of Appalachia, I am trying to do my part to care for this unique area's landscape and resources through managing the Berea College Forest. It's great to be able to be a part of something worthwhile with the College, and with the region as a whole. Maybe I have a chance to "become Appalachian", if not by birth, then by a common interest and solidarity with the region.

Southern Illinois, where I grew up, wasn't a bad place to live. But, it didn't have the strong sense of identity that Appalachia has. How many songs, books, or poems have you heard about rural Illinois, or even the Midwest? When I was a child, there were still a few "old timers" who butchered their own hogs, made quilts, did woodworking, played the fiddle, etc., but hardly anyone in the younger generations was picking it up or had an interest to… except for me when I was too young to help much or master all those skills. By the time I returned from college, almost all of these folks were gone. Most of the "old timers'" homes, along with their knowledge, just got bulldozed down to make room for a few more rows of corn.

Berea, on the other hand, seems to attract people who want to make a difference, learn new things, preserve traditional culture, and be creative. The College, by promoting for many years the traditional arts and crafts of Appalachia, has helped to stem the loss of these traditions.   The movement to preserve and promote Appalachian Crafts started about the same time Silas Mason came to Berea.

———————

It was a fine day for a shooting contest. Several "mountain men" had congregated together near the base of the mountain where those College folk came out each year to celebrate what they called "Mountain Day". The road through Narrow Gap was full of rocks and ruts, but it was only way through to Berea from the East…a good place to gather to see who was the best shot. Some of the men had heard that some new professor at the College had just bought some acreage in the vicinity and speculated as to what kind of a fella he was.

"I heard he's from up in New England", one remarked as his was lying prostrate on the ground…taking aim at the makeshift target with his sleek, long Kentucky rifle…"POW!" "Vermont, in particular", another added. "Shh— bang!"

"Went to college in Kansas, is what I heard", said another, as he got down on the ground and leveled a shot at the target. "Boom!"

"Well, what you suppose he's wantin' with a piece of mountain ground, anyway?" asked another mountain man as he reloaded his muzzleloader.

"Beats me", answered the first fella as he got up and pulled a piece of twist t'bacca out of his bib overall pocket, and cut off a piece. "There ain't much ground left to warsh off this mountain after growin' corn up to the cliff…ain't even hardly a squirrel could support himself on this ground anymore" he added as he slipped the piece of chaw into his mouth.

Silas heard the shooting as he walked his tall, black Tennessee walking horse down the road toward the group of mountain men. "Shooting contest", he said to himself as he debated whether or not turn around and go back to Berea. "No, I've got to get to know these men sooner or later, and this is the only nice weather I've had without a class to teach for a while. I've got to get out and look over that tract of land I bought and look it over and decide what to do with it."

The mountain men noticed a lone man on a horse approaching, and as they came closer they could see his "city clothes". One by one, they began to stand and look towards him. "Well, if it ain't that fella right there, the one we've been talking about", said one of the men who had recognized Mr. Mason.

Apprehension filled the air as Mason approached, cut his mind was full of his vision taking shape as he surveyed the landscape around him: "If only a large enough tract of land can be purchased out here in these cut-over, eroded hills for a managed forest capable of providing wood for the college in perpetuity could be achieved! Water from the springs could be piped into town so that the students don't have to roll barrels of creek water up the hills…wow, would they appreciate that! Mountain Day and year-around recreation could be enjoyed by our students and community, and forest management education and research opportunities would always have a home here!"

Silas hoped the men would be eager to get back to their shooting contest and let him quickly pass undisturbed. Waving at them as he stopped his horse a few yards from the string men, he offered a greeting and they returned a welcome. "Let's see what kinda shot you are, mister!" one of the young fellow excitedly beckoned, as he waved his rifle in the air. "Yeh!" chimed in several others, as one of the men, with a few rusty varnish can lids with a red dot painted in the centers ran to the big log in the distance and began to nail it on for a fresh target.

This is exactly what Mason did not want. He thought, at worst, he may be held up for a time with idle conversation…but not this! Not a shooting match! He had no experience with marksmanship, and felt that his best bet was to excuse himself from certain embarrassment. "Oh no. I thank you men, but I really must be on my…"

"Oh no, you ain't in such a big hurry as all that now. Just show us what you got, mister!" begged the first young fellow, while an elderly gentleman approached him and looked him squarely in the eye in a kindly but direct manner as he handed him his own, loaded, gun.   As Silas unthinkingly took hold of the gun in his own hand, the crowd approvingly nodded and the younger men shook their fists or whooped at him.

The men insisted he give it a try, and expected him to lay down on his belly as they had been doing. Instead, Mason hurriedly raised the gun to his shoulder and without changing his standing position at all, he fired off a round at the target. "Ker-pow!!"

Silas' shot knocked the center right out of the target! While the men stood in astonishment; staring at the rusty can lid with the red dot blown out of it, Silas quickly shoved the gun into the hands of the old gentleman and mounted his horse. "I'll never be able to repeat that lucky shot!" he thought, as he tried to appear unsurprised and fully composed…as if anything less than a bull's-eye would have been the exception for such a marksman as himself.

The mountain men around Berea had great respect for a good marksman and they readily accepted Silas Mason after that.

(This is my dramatized version of an account described in "Glimpses of Kentucky" by former Berea College Forester, John F. King).

I wonder how many times Mr. Mason recounted that story during his travels in Northern Africa and the Middle East after he left Berea to accept a position with US Department of Agriculture. His illustrious career did not end in Kentucky. He was instrumental in successfully establishing date palm cultivation in southern California. Indeed, his graduate work at Kansas studying date palm cultivars and his eventual career would involve researching various cultivars across Egypt, Libya and the surrounding regions and bringing them back to try in America. Perhaps this is a subject for another book!

———————

"I wonder if anyone is doing tatting?" my wife asked me as we walked past the gentleman turning wood on a foot-powered lathe and went in the log cabin at the Cradle of Forestry. On our honeymoon, we had wandered from Bays Mountain to Mt. Mitchell, then on to the Cradle of Forestry at Brevard, NC. Knowing she hoped to find time to resume her tatting and find a place for herself in Berea's arts and crafts community, I answered: "I don't know if anyone is doing that in Berea...maybe that could be 'your thing'".

"Berea!? Are you from Berea?!" a woman inside the log cabin remarked. I told her I was the Berea College Forester, and she thought we were cool to want to go there on there on our honeymoon. She went on to explain how much the Appalachian arts & crafts revival had to do with Berea College's efforts in the early 1900's and how much she loved Berea.

Berea College is not only home to one of the oldest continuously managed forests in the United States...but also is home to one of America's "arts and crafts capitals". From Berea College's own visitor's center and arts stores to the huge Artisan Center at the interstate, some of the most beautiful art can be observed and purchased. (https://www.visitberea.com/; http://www.kentuckyartisancenter.ky.gov/visit/)

It is no accident that Berea became known as "Kentucky's Arts Capital", as Berea College is largely responsible for what has been termed the Appalachian Handcraft Revival. The following is from Berea's College's website:

Berea College President William G. Frost, during the late 1890s, asked a local weaver whether she thought orders could be filled in a month or two for a half-dozen duplicates of a hand-woven coverlet made to his standards. Her reply, which Frost referred to as his "first lesson in weaving", summed up the difference between handcraft and manufacturing.

"President Frost, in order to make so many kivers we will have to raise more sheep, shear them, pick and wash the wool, card it and spin it, then collect the bark and such to color it. Then we will have to have the loom all set up, fix the warp and beam it, then get a draft and thread the warp for the pattern we want, then tie up the loom, and then we will be ready for the weaving . . . It would take we'uns nigh one year or more to afore we could have that many kivers wove. It's no child's play to weave a kiver, president."

Frost's "lessons" in weaving and other mountain crafts began in 1893, during an all-summer horseback ride through the mountains of Eastern Kentucky, West Virginia, Virginia, Tennessee, and North Carolina. The so-called "extension tour" took the Frosts into many mountain cabins where handwoven coverlets, baskets, chair bottoms, and rug were in daily use. But the "kivers" most attracted Frost, who purchased several examples for resale, and it was not long until coverlets became part of the medium for exchange at Berea College; coverlets bartered for educational expenses.

Frost saw, in the mountain coverlets, the potential for preserving a craft tradition and – at the same time – developing a new market for Appalachian crafts.

Berea's "Fireside Industries" began in 1883, followed by the "Homespun Fair" first staged on Commencement Day in 1896, and Berea College's history of leading the Appalachian craft revival was firmly entrenched. The city of Berea's US Post Office features a WPA mural depicting a typical Berea Commencement Day and Homespun Fair. In 1993, the Fair was recreated for a 100 year celebration.

By 1899, Berea's weaving and marketing effort required the full-time energies of Miss Josephine Robinson, the dean of women, and in 1900, a Berea coverlet won a gold medal at the Paris Exposition. By 1903, Berea's sales were up to a new high mark of $1,500 per year.

In 1911, Frost brought Mrs. Anna Ernberg, a native of Sweden, to Berea to direct the Fireside Industries. In his letter hiring Mrs. Ernberg, Dr. Frost spelled out what he expected: "We do not wish to introduce forms of weaving which are new and foreign to the people here but to encourage and develop the forms which have been handed down by tradition from the old English and Scottish sources."

Mrs. Ernberg's influence on the weaving of the Appalachian region lasts to this day. Among those who came to Berea to study were Lucy Morgan, founder of the Penland School for Crafts, and Lou Tate, founder of the "Little Loom House" in Louisville. Loom building patterns or Woodcraft-built hand looms were provided to the region. Weaving teachers from other settlement schools and craft centers came to learn both hand weaving and crafts marketing.

In 1926, Berea graduate Howard "Tony" Ford took on the job of creating a department to weave wool yardage on fly-shuttle looms. The "Mountain Weaver Boys," in addition to producing blankets and yardage for garments, took on jobs such as weaving 198 yards of fabric for the 1937 yearbook covers. The "Boys" continued through 1941.

Woodcraft evolved from the College's cabinetry program with design influences from famed furniture collector, artist, and author, Wallace M. Nutting, whose collection is on display on the second floor of the Log House Craft Gallery.

Needlecraft was added when the manager of the Boone Tavern Gift Shop put student clerks to work, between customers, sewing "cuddle toys". Broomcraft began when an obsolete broom factory was donated to the college. The Candy Kitchen, part of the Berea College Bakery, made decorated tea sugars for a national market.

What is now the Log House Gallery was built in 1917, when Mrs. Ernberg raised the funds for a new weaving workshop. At that time, only the second floor was used for a gift shop. Mrs. Ernberg then raised the money to build the "Sunshine Ballard Cottage", used for many years, including today, as the studio of Fireside Weaving. Since the early 1950s, the museum area has served as a primary outlet for Appalachian crafts including work not done at the college.

Craft leaders from centers in North Carolina joined with Kentucky and Tennessee centers to plan a regional cooperative effort that still serves the craftspeople of Appalachia.

During the late 1920s, Berea College played a major role in the planning for and formation of the Southern Highland Craft Guild, a nine-state membership craft organization dedicated to craft preservation, teaching, and marketing. In 1930, the Guild was formally organized with support from the Council of the Southern Mountains. Berea's fourth president William J. Hutchins was very much involved, and Helen Dingman, of Berea, chaired the first meeting at Penland School. She later wrote, "We met on a mountaintop literally, and the freedom and friendship of the group as they talked over the mountain handicraft situation – the hopes and fears and practical problems – made it real mountain top experience."

For many years, the Guild's records were maintained at Berea College. From the beginning Berea's faculty, staff, and students have been active in guild programs.

The early craft history of Berea College and of the Southern Highland Craft Guild is covered in Allen H. Eaton's classic book, Handicrafts of the Southern Highlands, first published in 1937. Berea's weaving programs are the subject of Philis Alvic's study, Weavers of Southern Highlands: Berea, part of a series on mountain weaving centers of the early 20th Century. Garry Barker's 1991 book, The Handcraft Revival in Southern Appalachia, 1930-1990, takes Eaton's study on into the modern era.

Throughout the first half of the 20th Century, Berea graduates were many of the region's weaving and woodworking instructors, the producing craftspeople, and the settlement school and agency staff members who helped create the solid markets for quality crafts made in Appalachia.

In 1960, Berea College helped sponsor the formation of the Kentucky Guild of Artists & Craftsmen. The college provided free office space and the two boxcars that traveled Kentucky as the Kentucky Guild Train, probably the nation's first mobile arts programming.

In 1970, Berea's ceramic department became the Ceramic Apprenticeship Program, a formal crossover between classroom and practical work experience, an opportunity that has trained dozens of young studio potters. The Wrought Iron Program was added during the early 1970s. During the early seventies the Candy Kitchen was closed, and 15 years later Needlecraft was no longer financially viable and was closed.

As Berea College shifted from a self-sustaining community, complete with college-operated farms, dairy, bakery, brick factory, hospital, and fire department, the "Fireside Industries" shifted more to crafts production for retail markets. The college and the Southern Highland Craft Guild created a market for quality regional crafts in the traditional sales centers of Berea, Gatlinburg, and Asheville, plus shipments to the entire nation.

Today almost 115 students and 13 full-time staff work in the Student Crafts Program, making and marketing the wood, brooms, weaving, jewelry, and pottery. Three shops in Berea – the Log House Craft Gallery, the Boone Tavern Gift Shop, and Student Crafts on the Square (SCOTS) – are student staffed.

Campus workshops are open to the public weekdays, and craft tours leave twice daily from the Boone Tavern Gift Shop. Weekend craft demonstrations in the Log House add to the availability, and College booths in select regional craft fairs continue the traditional Berea presence. Continued college involvement in craft organizations and marketing efforts such as the Southern Highland Craft Guild, the Kentucky Guild, and the Kentucky Crafts Marketing Program to carry on the work begun over 100 years ago during a summer horseback tour of the mountains.

The Student Crafts Program often provides a teaching laboratory for other college departments, such as Economics and Business, Industrial Arts and Technology, and Art. Recent studies have covered design for production, quality improvement, and production costs. Berea's unique work/study program, crafts production, and marketing efforts draw researchers from across the world to study current activities. Researchers studying the Appalachian craft revival come to Berea College's Archives and Special Collections and Museum to access over a century of collected information.

The most important aspects of the Student Crafts Program at Berea College continues to be the students who learn not just craft skills but a deep appreciation for the traditions and exacting standards of craft: quality materials, hand work, and expectations of excellence impact the lives of every student who spends time in the craft workshops of Berea College.

Berea's students, who come primarily from the Appalachian counties of nine states and who must demonstrate academic ability and financial need, do not receive academic credit for their work in the crafts. All Berea students work 10-20 hours a week in lieu of paying tuition; those who choose the crafts areas also earn the lifelong dividend of the handmade experience. It is still, today "no child's play" to weave a blanket, throw a vase, shape hardwoods into stools, baskets, and beds, or roll a broom from natural corn. https://www.berea.edu/student-crafts/history/

I think it is no coincidence that both the beginning of the arts and crafts revival at Berea and the beginning of the Berea College Forest happened together at this remarkable institution. The "art of forestry" has thrived amid this "arts and crafts" community simultaneously. I think the two go together in ways that are not yet fully discovered and articulated, even today. Currently, the Forestry Department is engaged with supplying sustainably horse-logged timber from the Berea College Forest to the Crafts Department and I hope to expand this to make horse-logged materials available to the larger community of woodcrafters as well.

Foresters are among of the most "place-based" professionals in the world. We necessarily have to recognize, learn about, and embrace our own place in the world; because the forests we manage are unique wherever we are practicing our profession. When I went to SIU-C to earn a forestry degree, I had some classes that provided an overview of forests in other regions of the United States…but the Central Hardwood Forest type is what I mainly studied because that is the forest type where I was at.

Sure, foresters can learn about more than one forest type and successfully manage different types of forests as they gain experience in different regions. However, as someone who grew up spending time in the forests near where I lived, I cannot imagine being a forester where there are no oak trees. Most foresters I know have a favorite type of tree, one that invokes feelings of "home" to them. Spreading, majestic, big old oak trees…especially white oak trees…are my favorite. The white oak also happens to be the state tree of Illinois. The stately, towering tulip tree…also known as yellow poplar…is the state tree of Kentucky, and is also a fantastic tree which also is the main feature of the symbol that the Berea College Forestry Department adopted to represent our program. Southern Illinois is unique in that it lies near the convergence of the extent of many tree species' native ranges. Bald cypress, *Taxodium distichum*, swamps… which look a lot like what you would find in Louisiana bayous…are found in the Cache River of Southern Illinois, just a few miles from where I went to college…while Northern Red Oak, *Quercus rubra*, and other more northern species reach down that far as well. A little piece of the Ozarks to the west reaches across the Mississippi in the Pine Hills, where shortleaf pine, *Pinus echinata*, is found, while yellow poplar, *Liriodendron tulipifera*, and American beech, *Fagus grandifolia*, and even American chestnut, *Castanea dentata*, reach all the way over from the eastern hardwood forests.

But, the forests of Appalachia are even more diverse than those of Southern Illinois. In fact, the southern Appalachians are one of the most diverse temperate forests in the world. Elevation changes mimic changes in latitude, and many northern "relic" species of trees, plants, and wildlife that lived here during the ice ages remain in scattered high elevation locations even today. Nearly 10,000 species of organisms are known to inhabit Southern Appalachia, and more species are still being discovered on a regular basis. Salamanders and fungi, in particular, reach their highest levels of diversity here, and many of these species are endemic: found nowhere else in the world (http://highlandsbiological.org/nature-center/biodiversity-of-the-southern-appalachians).

Wherever you find yourself, your own unique place in this world has its own unique ecosystem, its own diverse arrangement of species and opportunities. Most states offer assistance to landowners through a state forestry program and/or a state university extension service. Generally, these services are free, and landowners can obtain a forest management plan to help guide them in the management of their forests. While I was a District Forester for the Illinois Department of Natural Resources, this was my job. I would encourage all landowners to participate in these opportunities that are available. Forestry consultants, who operate their own private businesses to provide assistance to landowners for a fee, are also a good option. Typically, the free advice given through state agencies and university extension services only go so far…and forest owners who wish to have timber sales and other extensive assistance will find that a forestry consultant is well worth the fees charged.

I would encourage landowners, however, to also educate themselves and become as much a forester as they can, in their own right. Able-bodied individuals, especially, and those with children and grandchildren, or other extended family who enjoy the forest, can make forest management fun and rewarding. Private land ownership is a treasure of opportunity. Those who own land…

especially forest land…have the opportunity, and perhaps the responsibility, to send their "postage stamp" parcel of land on this planet into the future for subsequent generations to benefit from.

When I became a District Forester, I was just as often learning from landowners as I was educating them. One of the most rewarding parts of my job was visiting with and learning from "old timers". I feel very fortunate to have been a service forester in time to interact mainly with the "greatest generation", the World War II generation, early in my career.

During my first week on the job, I perused through the file cabinets to find the oldest forest management plans I could find. Computer-generated plans had just become the "in thing", and most of the files were full of older plans that had been hand-typed…typically about four pages long as opposed to the 20-something or more pages of the newer plans. What I found was infinitely interesting. These very old plans…from the 50's and 60's… did not go into much detail. They assumed landowners…most of whom were farmers…already knew their tree species, the uses for the various types of wood, and knew how to harvest trees, saw them up, and use them. Advice given was mainly guidance for long-range planning. Old brochures showed photos of farmers logging with horses, mules, and tractors. Lists of sawmills were by county, and there were a lot of them…usually there'd be one within ten miles.

"Wow", I thought. There was a "local economy" back then. People knew how to do so much by themselves, and their communities were practically self-sufficient. Every little "burg", as my grandma would call them, had a grain elevator, sawmill, grocery store with a butcher shop, lumber yard, service station, hardware store, car dealership, and so on. Even towns with as few as 400 people had all this…as well as their own community schools. Mostly, after local needs were met locally, only the "excess" was sold elsewhere.

In the 90's, in my early days as a service forester, the land was mainly owned by the "old timers" who had lived this way. They might have been in their 70's and 80's, but when I recommended a "timber stand improvement" they nearly always did their own work. One fellow, 83 years old and with a fused back that prevented him from bending over, started his maple midstory removal project (cutting literally hundreds of maple saplings per acre across his 80 acres of land) the very day after he received his management plan. His wife treated the cut stumps with glyphosate herbicide as he cut them. They worked sunup to sundown six days a week until it was done! Woods owners could sign up for "cost share" assistance, which meant that the government would pay them to do this, but many of these landowners did it without the payment!

One of the many inspiring landowners I met while I worked as a service forester was a man who called me and asked if I knew anything about drying pine lumber. He said he had sawed enough pine lumber off his land to build a house, and wondered if I could advise him on how to dry it properly. "What?! PINE lumber?" I thought to myself. Pine was not native to my part of the state, and pine trees in quantity and size to saw into lumber were unheard of.

I asked him where he came up with the pine, and he told me how his dad had followed the advice of the "conservation service" in the 50's and planted their marginal, eroded land to white pine and loblolly pine. He was ten years old at the time, and his dad told him these were his trees…that someday they would build a house together with them on the family farm. By golly, it was time to do it. The trees were now 18 to 22 inches diameter and straight as an arrow. They had cut what they needed, sawed them up, from some of the lumber built a shed to store them in, and wanted to make sure they dried them right.

I had the opportunity to visit the farms of many such remarkably resourceful folks. One farmer showed me the sawmill he had built; with the help of another relative, out of salvaged combine parts. The power source for the mill was an old Trans Am engine from a salvage yard! It left me wondering who the "experts" here were. Not me. As the World War II generation began to pass away and pass on the land to the "baby boom" generation, this "can do" philosophy was largely lost. About this same time, the family-owned businesses disappeared and were replaced by nothing but a convenience store at first…and then a Dollar General later. Now, hardly any of these little "burgs" have anything going on at all.

The only thing remotely close to resembling a community in the rural areas now is the Amish communities. Fortunately, to my way of thinking, my "old stomping grounds" where I "traipsed around" in the creeks and forests as a child, have been largely bought up and colonized by the Amish. Just before my grandmother passed away, I took her for a drive through this area and she remarked that it looked like "home" again…the way she remembered it as a little girl. We passed Amish couples in their horse-drawn buggies, and on their bicycles, as we drove by farms that had been owned by "old timers", she grew up with whose heirs had "sold out".

The old cemetery that was along Flat Creek, where she grew up, was now within view of a huge, beautiful Amish home and farm that operated a huge greenhouse. A little girl, about three years old, was in the garden picking green beans and placing them in a basket. We got to see a large group of Amish moving a huge old abandoned home that was given to them, about eight miles across fields with many teams of draft horses. "That house used to be in Shobonier", Grandma remarked. "I knew the people that lived in it. It used to be the prettiest house in town. We used to go to Shobonier a few times a year to get supplies. There used to be several stores, a blacksmith shop, and a school. Nothing there now. The whole town is nothing but a wreck now, gone to the dogs…sure is good to see somebody making a go of things now."

The house did turn out to be beautiful again. They placed it on a basement they had already built, I found out later, on a tract of land I had hoped to buy myself when I was 18 but had no money…no community. I just saw, last month as I returned to my home area to attend my family reunion; that the house and sixty acres is for sale. The Amish family who moved and restored the house are now selling it to move out of state. I wish I could buy it…but would be reluctant to move back to Illinois now. It has become so high-tax that people and opportunity are leaving that states in droves, and even after one governor after the next ends up in prison, nothing has changed. I hope enough Amish decide to stay to keep the little Amish store open nearby that has the best soft pretzels you can find, and about anything else you might need.

My point to make in this little story is that yes, we should consult professionals. As a forester, I urge landowners to seek advice and services that are available. But, things have become so consolidated, so global economically, that even if landowners do this and seek bids for a timber sale and so on… there are fewer local mills left to sell to…and the loggers that are still in business have likely had to "get big or get out". This makes small tracts not feasible to manage unless we can find new ways to do it. Government programs can come and go…and sometimes when a tax is put in place to make one go, the program goes away but the tax stays.

To "make a go of it", we may have to get it going ourselves. Every place is different, with a different flavor, a unique culture, a unique and diverse ecosystem of plants and animals. Yet, every town now has all the same franchises, signs, supply chains running all the way to China…while we have most of what we really need right here already. The sun, the water, the soil are the basis for everything we have, and the people that do the labor to utilize it is our greatest resource…yet they are not cherished as they should be in present economic and social system. Unique places, people and natural resources are all treated the same, everywhere…cheaply and disposably.

Wendell Berry said it best: "What I stand for is what I stand on". The soil. It is washing away and blowing away, just as our rural communities are doing. I had the opportunity to discuss these matters a few months ago with Wendell Berry while I was at a "Steep Ground Logging" training session near Russell Springs, Kentucky. It was an honor to just be around Wendell and listen to him. The logging was being done with horses and mules, and was an annual "Biological Woodsman Week" put on by the Healing Harvest Foundation. Wendell and I lamented how, even with such services offered by University Extension and State agencies, the condition of our environment and our communities mostly keeps getting worse.

Wendell gave one of his inspiring talks to us over dinner one evening in the basement of a rural church near the logging site after a full day of "worst first" logging. "Worst first" logging is where the worst trees are taken out first, to improve the health of the forest…leaving the best trees to fully mature before being gradually harvested sustainably, as they are replaced naturally. When critics of this type of management invariably remark that doing it so slow this way; especially with horses, that we will "never get done"… Wendell invariably answers "Isn't that the point? We don't want to ever get done; we want to do this sustainably, in perpetuity."

Perhaps it was fitting that Wendell addressed our bunch in the basement of a church. We were eating a delicious meal prepared by our host, Ben Burgess' wife and mother. There was home-made sorghum molasses syrup made by Ben and his dad and uncle. I had watched them squeeze the sorghum the week before with an antique press that that they had mounted to a treadmill that one of their horses turned by walking on the treadmill. We stood around and stirred the juiced for hours until it was cooked down to delicious syrup. There was "chicken of the woods" mushrooms his family members had found on logs in the forest there in a bowl to put on our salads and home-made squash pies.

Wendell liked what we were doing, and wanted to come down and support it. The family farm we worked on, the community, the little rural church… were important places to all of us….sacred places, really. Not places to trash, to treat simply as a commodity, an investment, as raw materials. Perhaps Wendell said it best in "Give": "There are no unsacred places; there are only sacred places and desecrated places."

———————

Today's "data driven world" seems obsessed with collecting information. I have a natural aversion to taking the time to collect data; to do an inventory, before plunging into whatever I have in mind to accomplish. But, as much as I like to emphasize the "art" or "craft" of doing things, we must make sure we have the supplies we need. Forest management is no different. An inventory of the forest…especially a large one…is necessary for making informed decisions on a long term basis.

While attending forest measurement classes at Southern Illinois University, I was particularly impressed with the science behind "point sampling". "How could you accurately inventory a forest by randomly looking through this glass gadget at random points and recording what you see through the instrument?" I thought to myself. "Whoever came up with this must have been really smart."

To learn forest inventory, we had to verify our sampling with a total inventory. That is, we would measure all of the trees in a given area that had been sampled, to find out if our sample was correct. "Wow, it really does work!" I soon realized. "Point sampling" or the "Bitterlich method", is regarded by many foresters as the "invention of the century", as it enables forests to be inventoried accurately by sampling rather than counting and measuring every tree.

Being a history nut, I naturally had to find out more about the inventor of point sampling; Walter Bitterlich. To my astonishment, he was still living and still inventing. My first web search of him revealed pictures of him ice skating on his 95th birthday. I just recently discovered that Mr. Bitterlich finally passed away a few years ago. He died ten days short of his 100th birthday.

Walter Bittlerlich grew up in the mountains of Austria, where he was a fourth generation forester. His professional life was interrupted when he was drafted by Germany to serve on the Russian Front during World War II. During the war, he was instrumental in perfecting the design for what was to become the modern snowmobile. After the war, Bitterlich was able to resume his forestry career and soon patented his "Winkelzaehlprobe" or "angle-count sampling" instrument, called the Spiegel-Relaskop. These instruments are still state-of-the-art and used worldwide for forest inventory and other applications.

In 1966, Dr. Bitterlich was appointed professor at the University of Agriculture in Vienna, Austria, where he was a forestry instructor until his retirement in 1978. He continued to invent, and he holds many patents on innovations ranging from snowmobiles to garden tools and furniture-manufacturing equipment. Bitterlich participated in nearly all World Forestry Congresses and Symposiums throughout his life. He was also known as a very humble and gracious host. Even after achieving fame and fortune, he continued to live in his modest home in Salzburg where he graciously received the many foresters who would make a pilgrimage to Austria to meet him. I never was able to make that pilgrimage to Salzburg; but Bitterlich's example of forestry plus technology has served as an important example to follow in today's age of computers, GIS, GPS, etc...

"Knock, knock!" Someone was knocking on my door early in the morning. I got up, put on a robe, went to the door and looked out the window. There was a big, bearded, burly guy standing there with an angry look on his face. I recognized him as one of the loggers I had been supervising on a job I had marked and inventoried. They had just finished it up the week before and had been paid. As soon as I went outside to talk to him, two other big burly men stepped out from behind the corner of my house. I could see they had their fists clenched. My first thought was to look around for something to use as a weapon. The first guy bellowed in an angry voice, "Where's the rest of our _____ money?!"

I couldn't understand why he was upset. That particular job had "cut out" to almost the exact footage (volume) I had estimated it would be. It had actually over-ran by a few hundred board feet. I had felt a sense of accomplishment in having so accurately scaled that job...and yet these loggers were threatening to beat me up because they felt cheated that there wasn't much "over-run". When they realized I wasn't going to give them any more money, and didn't act afraid of them, they cussed me out and left. I ended up finding another logger to replace their operation...one guy who worked alone but could outwork their three-man crew by himself. Nelson Bishop was in his 60's and yet would work from sunup to sundown and in any weather...with no wasted motion or error. One day, I noticed he often brought no lunch and worked all day with nothing but a jug of water tied to his skidder. When I remarked about this, he just said "Everything is easy after you've worked nine years in Prudhoe Bay, Alaska on oil derricks in nearly 100-below-zero wind chill." He had saved his money up there and came back to Illinois and bought some land and a skidder and started his own business. Smart.

Nothing feels worse than being cheated. Many landowners have been cheated, and loggers too. The most notorious timber thieves I have known over the years were so slick that hardly anyone they preyed upon realized they were cheated. Conversely, many of the most honest and well-meaning folks in the business get falsely accused of cheating people. Despite their commonly held bad reputation, some of the most hard-working and honest people I know are in the logging, timber and sawmill business. However, timber value is largely an unknown to most; and once the trees are cut, it is even more difficult to know what happened or be able to prove something one way or another. From that day on, I knew I was capable of performing a very accurate inventory...but I always lowered the figure I gave out by about 15% so that the loggers and landowners would feel like they came out on top...like they got something "extra". I also took more precautions before answering my door!

No one came to me, as a District Forester, looking to manage their timber. Well, almost no one. Nearly everyone who contacted me did so to receive cost-free assistance to have a timber sale. Most had already been approached by a timber buyer or logger and this is what prompted them to look for help. They feared being cheated or having their timber ruined. Unfortunately, there are pretty much just three options most owners perceived for their forests: 1) don't cut anything 2) cut everything 3) cut the trees that are selected by the timber buyer offering to buy their timber.

"Selective" harvests usually result in the trees larger than a certain diameter…usually all the most valuable trees… being selected to cut. This practice is referred to as "high grading", and results in a severely degraded forest. I often refer to it as "cut the best, leave the rest". It is the opposite of good forestry. I like start out with "cut the worst first".

Foresters, when deciding on a harvest method for any particular forest stand, evaluate many factors such as markets, tree size, species, stocking (a measure of the area of space occupied by trees, usually in square feet per acre), spacing, soil type, and insect and disease presence to determine management options for a particular area of forest. Whatever the forester surmises should be done, the landowner makes the ultimate decision. Quite often, they turn right around and do what the logger proposed in the first place. This can be very frustrating for the forester. We may invest a lot of time and effort performing an inventory and writing a management plan only to have the landowner turn to an unscrupulous timber buyer to do the opposite. This happened so many times during my career that I became pretty cynical about the whole "landowner assistance" programs and hoped to find a stable place to manage one large forest rather than keep trying to convince multiple landowners to manage theirs.

Silviculture is the word which is used to describe the various techniques used to manage forests; and comes from the Latin silva (forest), and culture (cultivation). Silviculture on the Berea College Forest is focused on producing large, high-quality trees, keeping high canopy forest on the majority of the acreage at any given time, and maintaining spacial variation of different stand ages as well as physical and compositional structures that support native biodiversity. Silvicultural practices commonly used for timber harvesting on the Berea College Forest include commercial thinning, shelterwood, and small "patch cuts" or group openings for regeneration purposes.

I don't think anyone likes to see forests harvested; especially if they are attached to that "place." At least for a while, harvested areas look messy. Best Management Practices (BMP's) for logging…the approved way of doing things by the state…are mandatory in Kentucky, but skid trails can still look bad until grassy cover is well established on them. A well planned and implemented harvest may, in fact, not look better to the untrained eye than an improper "high grade" harvest.

Loggers, timber buyers, and even foresters are typically blamed for the poor condition of many of our woodlands. Sometimes, forests really are mismanaged, while other times this is just a matter of perception. While there are some unscrupulous timber buyers, loggers and foresters (as with any profession), I have found that most are honest, hard-working people just trying to make a living providing the wood that our society needs. Ultimately, it is the landowner who is responsible for how their forest is managed, mismanaged, or not managed. This is another reason why a pervasive, culture-wide "land ethic" is so important.

The Berea College Forest is managed for a variety of purposes; including for timber production. When Dr. Silas Mason established the forestry program over one hundred years ago, he envisioned the forest providing a continuous source of timber for furniture production, building material, and income. He also envisioned it supplying water to the college from its many springs and, later, reservoirs…as well as educational opportunity and recreation.

In the early days, logging was done with horses and mules. I have several historic photos of horses hauling timber out of the Berea College Forest. Our contract loggers have found old horse shoes and logging "grabs" from previous harvests which proves this. Eventually, production was mechanized, and Berea College even operated their own logging and sawmilling enterprise for quite a number of years.

Despite all the harvesting over the years, the Berea College Forest now contains far more timber volume than it did when the property was purchased. In fact, much of the property was, when the College purchased it, highly eroded "worn out" farm ground and heavily cut-over forest. Historic photos attest to this fact; showing severely abused, eroded and nearly treeless views of the Pinnacles (our beautiful rock cliffs) and beyond.

Some people think forests should only be "preserved" and never harvested. However, a forest is dynamic and always growing. As trees become larger, they become crowded and some die. Some blow down, and others are damaged from insect and disease outbreaks. Forests can also, in this day and age, be overrun by invasive species. Consequently, informed decisions and appropriate actions by foresters can actually improve the forest through management which includes harvesting.

Mankind has, in one way or another, managed the landscape since the earliest of times. Ancient civilizations and aboriginal cultures managed it largely through the use of fire. Few now realize how extensively Native Americans manipulated their environment through burning, cultivation, and clearing.

Although the Berea College Forest is a managed forest, this does not mean every acre is open for harvesting. Nearly one quarter of the forest is designated as "Legacy Stands". These are areas we designated as having special qualities that merit leaving them to develop without harvesting… buffer zones, scenic areas and other unique places which are off limits to harvesting unless they would be salvaged as a result of severe damage by fire, tornado, etc…

To me, this all makes good "horse sense". However, at the same time, I am always looking for ways to utilize the forest resource in a more sustainable, low-impact way. This is why I contracted some mule-loggers to harvest fire-damaged trees on steep slopes that ran directly into our main was supply: Lake Owsley. The use of conventional equipment on these steep slopes so close to the lake would have been difficult to use without the potential for causing erosion and siltation.

———————

Fred didn't look big enough or strong enough to pull the huge red oak butt log one foot; let alone a quarter mile. I was afraid it would start rolling on the steep hillside and jerk him down the slope too. That would be bad…real bad. I had watched horse and mule logging before, in Illinois…even helped with it. But, this was different. It was STEEP here! Fred needed to pull it into position by himself, as there was not enough room where the log was at to employ a team of two just yet. Fred seemed to know what the plan was, and he just calmly waited for his instructions from Glen West. Then, he buckled down and all of a sudden his muscles tensed up under that sleek black/brown coat of his, and he moved the log. Over and over. He didn't care what other mule he was matched with or if it was Glen or Terry holding the reins. I even got to hold the reins a couple times. Shoot, he still did his job even if the fella behind him let go of the reins. He'd take that log to the log landing and wait for him to catch up.

Fred is the star of a video showing mule-logging for our Deep Green Residence Hall. Seven years later, he found his way back to Berea with a different owner and got to be in a video where he drags logs for the Western Flyer project…along with some big Suffolk punch horses. Now, he is helping teach students in the art of sustainable farming and logging for the Wendell Berry Farming Program of Sterling College.

Berea College has been in consultation a couple of times over the last several years with Jason Rutledge; a well-known horse logger from Virginia. This association has been like a long-dreamed reality coming to pass for me. I bought the July/August 1987 issue of "Mother Earth News" the summer I turned 17 and was applying to attend colleges. It hangs, framed, on my office wall. Jason and his horses are featured on the cover of that magazine, and reading his article at such a time was influential in me sticking with pursuing a career in forestry. Since then, I have hoped to be in a position to see sustainable forestry with horses utilized somewhere on a scale that could be appreciated and accepted.

Jason knows horses about as well as anyone, his preferred breed is the Suffolk Punch horse that originated in England. He knows forestry, and he has a "silver tongue" to talk about it to anyone who will listen. One of the goals of his foundation that particularly speaks to me, is the goal to: "Perpetuate truly sustainable cultural traditions that produce social capital through development of livelihoods that reward ground-level forest stewardship skills, protection of the environment, and preservation of family integrity." This goal is a truly worthy one, right? It speaks to the importance of culture, of people, of sustainably managing and protecting our environment. Well said. To learn more about Jason's ideas and methods, check out his website: www.healingharvestforestfoundation.org

Jason Rutledge trained Ben Burgess, who hosted the Biological Woodsman Week at his farm in Kentucky, and Ben has been training our own Berea College-owned Suffolk Punch team. Once our forestry staff and I become adept at using them ourselves, we plan to do demonstrations and small-scale harvesting on the College Forest with them. The horses were purchased and used for a short time by our former Forestry Technician, Bob Warren.

Bob was a gentle soul and ingenious man who had done everything from working in a steel mill, doing the accounting and bartending for a major ski resort "out West", skiing and riding motorcycles all over the West, working as a helitack forest fire fighter, earning a Ph.D. degree in microbiology, researching insects in Australia, building wooden boats, owning and operating an ecological consulting business, working for free for an Indian Reservation, teaching himself GPS and GIS like a pro, and finally working in forestry at Berea College after his wife, Wendy, came here to teach. While at Berea, he found his passion in Suffolk Punch draft horses and bought a pair.

Sadly, before Bob could fully realize his dream of starting a horse program at Berea College, he was struck by a fast-acting cancer and passed away. He did everything with passion though, and one of his last projects was to work with a group of "at risk" Berea College students to help them cope with college and life in general. He used Hanna and Holly, his gentle giants, to harvest several trees from the Berea College Forest with these students and together they built a small timber frame structure on the hill behind his and Wendy's residence on the forest. The success of his work with these students was evident when they showed up to his "celebration of life" gathering and each told of the impact he had had on their lives, as their eyes flowed with tears. Several of them related how Bob was the only "father figure" they had had, and how they couldn't believe someone cared so much about them, knew so much, had done so much in his life, and even how he handled his own approaching death with the same calm and gentle spirit as he had handled his life.

Perhaps part of the impact Bob made, and his ability to calmly handle the last chapter of his life the way he did, can partly be attributed to the presence of his horses.

Bob's life lives on at Berea College, through the College's horse program. What he initiated, on his own, has been adopted by the College. His horses were purchased by the College and while Hanna has since passed away, Holly and Willow [Hannah's replacement] remain here to be utilized on the forest, and with the students. His wife, Wendy, is the Coordinator for Berea College's Forest Outreach Center, and she gets to work every day within view of a display which features Bob driving Hanna and Holly, with the big grin on his face that he always had while working with them.

I think Bob exemplified…perhaps more so than anyone I have known… that *"No matter how much one may love the world as a whole, one can live fully in it only by living responsibly in some small part of it."* Bob loved the world and did something about it in whatever place he found himself.

———————

That small part of the world where we can show our love for the world as a whole by living responsibly, is our place. When I use the word "place", I can't help but think of the how some people have been "put in their place" or "kept in their place", and I would like to speak to that. While I portray in my forestry examples mostly men…white men…especially German men…this is because that is the history of forestry as a profession. I see the German model for forestry as the seed that bore fruit that has continued to our day. There are other, lesser known "heirloom" varieties of this seed to be discovered among indigenous cultures and there are new hybrids we can cross-pollinate with this seed to produce place-based versions of forestry.

America's "melting pot" has provided a place for much cross-pollination of ideas, cultures, arts and the sciences. One of the most famous American agricultural experts was George Washington Carver…an African-American slave boy born near the end of the Civil War. His father died before he was born, while he and his mother were kidnapped from his mother's "owner", Moses Carver, by bushwhackers when he was a baby. A neighbor chased after the bushwhackers, but was only able to recover George…trading Moses Carver's $300 race horse for him. Thus, he was raised by German immigrants; "free-thinker" anti-slavery immigrants who nevertheless purchased his mother to have help on their farm. He especially faced racism after left home and sought an education…believing as he would famously tell his students later, when he was an educator himself at Tuskegee Institute: "Education is the key to unlock the golden door of freedom."

Carver was initially turned away from entry into colleges because of his race. Eventually, he was admitted to Simpson College, a small Methodist school in Indianola, Iowa, that admitted all qualified applicants, regardless of race or ethnicity. Carver himself was an intensely devoted Christian and believed that God had given him a special gift and frequent visions to help him understand nature's secrets to help his people. He spoke often in such terms as: "I love to think of nature as an unlimited broadcasting station, through which God speaks to us every hour, if we will only tune in." He also held onto the premise that evil…including racism…as man's inability to "grasp the good".

As much as George Washington Carver loved nature, he also loved art. In fact, his first passion was to become an artist. He even won an Honorable Mention award for his painting "Yucca and Cactus" (of course it was a painting of plants!) at the 1893 Chicago World's Fair. However, he decided he could help his people more by studying his other interest…nature, plants, agriculture…and furthered his education at Iowa State. Studying agriculture there, where he was the first black student and was not allowed to eat or sleep with the other students…yet he, a black man, was president of the German Club.

(https://www.acs.org/content/acs/en/education/whatischemistry/landmarks/carver.html)

Race should never serve as a dividing line. Racism, sexism, and all the "isms" that divide us only separate us from loving one another. Love, I believe, is key to unleashing the creative forces that can, and should be, at work in all of us. To quote Carver again: "Anything will give up its secrets if you love it enough. Not only have I found that when I talk to the little flower or to the little peanut they will give up their secrets, but I have found that when I silently commune with people they give up their secrets also – if you love them enough."

Beautiful works of art are made up of a plethora of colors and forms. I believe that as art is incorporated more fully into the practice of forestry, it will be essential that forestry become a place for everyone to create new and beautiful solutions and works. Men and women of every creed, color, and nationality have unique backgrounds, perspectives, skills and imaginations that…like a healthy and resilient ecosystem…are essential for us to progress and create new works of art on this landscape; this patchwork of places we call Earth…our home.

At Berea College, I have had the opportunity to work with students from all over the world and from every walk of life. I am amazed at the diversity of perspectives and insights that they bring to my work and life experience. No matter how much we know, or think we know, we can always learn more…and we can all learn from anyone. Yet, for us to all function productively, we must form community…we must have some things in common.

I believe there is a "One-ness" that comes when two or more people; through common effort, understanding, and passion, accomplish something that they could not do by themselves alone. We grow as an individual when we realize that we can accomplish more together…and this is essence of communi- ty…and in church. Creativity also benefits, I believe, when we work inclusively. When we are all allowed to explore, learn, and apply ourselves to whatever "place" we feel inclined, talented, and motivated to apply ourselves, we all benefit.

Order and diversity may seem like opposing forces…and they can be, if there is not a common purpose. The word "Balkanization" comes from the Balkans…a mountain region with a lot of diverse groups and a long history of war between them that even turned into World Wars. Sometimes, it appears that the United States is heading in that direction too…

Isolation caused the diversity within our human race in much the same way that it created diversity among the other species; especially plants. Mountain ranges like the Appalachians contain so much diversity of plant life because organisms become isolated in this terrain and become adapted to the vast array of microclimates created by the elevation, moisture and climate differences.

People developed specialized adaptations and behaviors based on their environments but they are not plants…and no one should be assumed any longer to have a limited "place" where they must be content to be confined. There is a lot of accumulated value in this diversity…but also potential for harm when unexamined practices are implemented on a "one-size-fits-all" basis. One simple example is the green lawns developed in the French and English monar- chy credited for creating the cultural expectation in America for large, green lawns. The pesticides and mowing required in modern times to maintain them is quite harmful to the environment.

This Earth is our home. It's "all y'all's", as this Kentucky transplant likes to put it…and "home is where the heart is." As we progress as a society and as a civilization, to the point where everywhere is everyone's home, so will our hearts be fully employed to make it a true home…functional and practical, yes…but also comforting, inviting…beautiful. Everywhere. Later, I will introduce this idea of Order vs Diversity in the scientific terms of "Grouping Order" vs "Symmetry Order". First, however, we have to learn to think like a mountain.

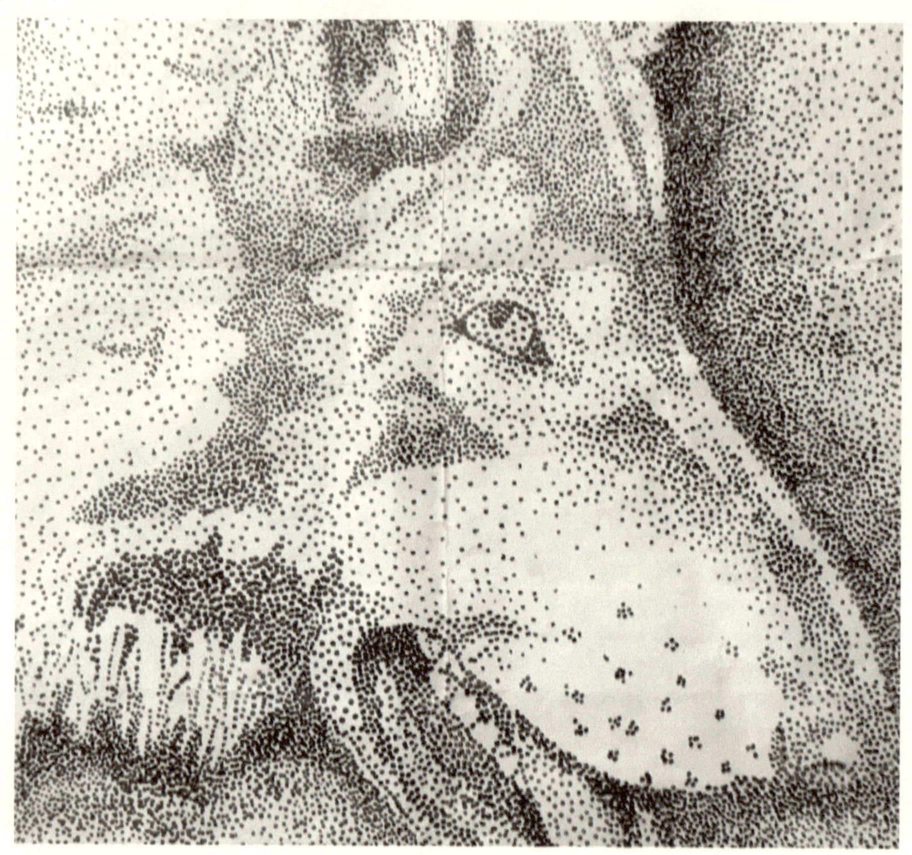

Chapter 3:
The Mountain

"Think Like a Mountain" – Aldo Leopold

"*We abuse land because we regard it as a commodity belonging to us. When we see land as a community to which we belong, we may begin to use it with love and respect.*" Aldo Leopold. Foreword, *A Sand County Almanac*.

Chances are, if you see me during the work week, I've probably got a shirt on that says, on the back, "Think Like a Mountain". This saying, by Aldo Leopold, the "Father of Ecology" and a forester, is what I have adopted as the Berea College Forestry Department's motto.

So, what does it mean to "think like a mountain"? The saying comes from the title of a poem, written by Leopold in 1949, as part of his book "A Sand

County Almanac". The poem laments the author's shooting of a wolf before he understood what the demise of wolves meant for the land. It then describes the consequences of eliminating wolves from the ecosystem; namely, the over-population of deer and subsequent destruction of the plants on which deer feed.

In a nutshell, Leopold realized that the same ecosystem which supports the wolf also supports the deer, the cattle, and humans. So, when humans take away parts of the ecosystem, a domino effect happens and impacts us all. The end result is then, as Leopold states, "…we have dustbowls and rivers washing the future into the sea".

While in college, I had to read "A Sand County Almanac", and became a fan of Leopold. I had a T-shirt back then with "Think Like a Mountain" on it and an abstract picture of a wolf howling, and a mountain in the background. So, for a long time, I've been considering what it means to "think like a mountain"; at least in relation to working in forestry.

I wonder what it would be like to live in an ecosystem which is intact with all its predators? Cougars are repopulating the Midwest. Wolves are spreading from Minnesota and upper Michigan. Bears are spreading too. If we're "thinking like a mountain", these occurrences should excite us rather than scare us, right?

Predators are the next step to restoring our native ecosystems, and they're on their way. Once an "apex predator"…a predator that was at the top of the food chain…is reestablished into an ecosystem, a "trophic cascade" effect happens and the whole ecosystem changes in many ways that scientists are only beginning to understand. The story of how wolves changed the rivers in Yellowstone National Park is a great example of this. The re-introduction of wolves changed the behavior patterns of elk and deer in the parks…keeping them on the move and in cover…they could no longer over-browse the open areas along the rivers. This enabled the trees to grow back along the waterways…which enabled the beaver to become reestablished and feed on these trees, dam the rivers, and create wetlands. The whole ecosystem was revived by adding wolves back into it. Leopold's epiphany came as he watched the "green fire" go out of the dying wolf's eyes after he shot it and began to contemplate what the absence of wolves would mean. Now we know for sure… (https://www.yellowstonepark.com/things-to-do/wolf-reintroduction-changes-ecosystem)

Isle Royale (1992). Fine time for my flashlight battery to begin to lose charge. I didn't have any more batteries along on the trip, so I decided not to turn it on as I quietly exited the tent and cautiously walked slowly down the trail to use the outdoor toilet a short distance from our camp site. The night air was cool and quiet and it was too dark to see anything except a faint silhouette of the tree tops against the night sky. Shuffling along and feeling the packed trail with my feet, I suddenly felt that the darkness had somehow become even darker… pitch dark. The hair on my neck stood up as I stopped in my tracks and felt for the flashlight button. This seemed like an emergency worthy of using the last trickle of my flashlight batteries to take a look around. As I pointed the flashlight in front of me and turned it on, I saw a wall of fur in front of me and my first thought was "sasquatch!"

In an adrenaline-filled burst of motion, I felt myself automatically leap to my right…behind I tree I didn't even know I was next to. As I leapt, I felt a brush of fur against my forehead. The large, furry wall exhaled as I tilted the flashlight upward and saw that "Bigfoot" was a bull moose! The moose grunted as the light shined in his eye, and I quickly tipped the beam downward. He had stopped just as I had; barely one step before we would have collided into each other. His beard is what brushed my forehead. As he began to resume his walk down the trail, I shined the flashlight at the moving wall of fur and timidly reached out and very lightly touched his fur as he went by. He calmly kept walking. Somehow, it seemed that the moose had greeted me on the path by stopping and just calmly walking on after I moved out of his way.

Isle Royale was especially known for its wolves, and I couldn't help but think of them in terms described by Aldo Leopold. I didn't get to see any wolves on the trip, and have still never seen wolves in the wild. But, on the day I got to know my future wife at Bays Mountain, TN…where we would get married several months later…we did get to look into the eyes of wolves in a remarkable experience. As if to show us that they approved of our relationship, as we walked past the wolf pen all of the wolves lined up in single file in front of us. They looked us in the eye and then the black one…the female alpha wolf… turned and began to walk in a circle. The others followed one-by-one, in single file and they ran in a big circle as if performing a circus act. Then, they jumped onto the wooden platform in the center of their large enclosure. Standing in a perfect row, evenly spaced, they stood there for about a minute looked right at us again. Maybe it is just a coincidence, but animal behavior sometimes seems to acknowledge us when we are "on the right path".

The trip to Isle Royale was during my last semester of college. All I had left to complete before I graduated was my summer course: a trip around Lake Superior. We went up the "mitten" of Michigan, crossed into Canada at Sault Ste. Marie, camped in Pukaskwa National Park in Ontario, canoed several days in the Boundary Waters and visited Isle Royale National Park. The forests were awesome. It was the first time I travelled outside the United States. We saw bald eagles and loons in abundance, and black bear snooped around our tents at night. I caught a big pike on my first cast…the only fish I caught on the trip… and it made a nice meal over the campfire. But, I was expecting to see some cultural diversity too…and didn't.

"Diversity" is a word commonly used today. We hear it all the time. We celebrate cultural diversity, ethnic diversity, racial diversity, and all manner of diversity…in our demographics. The same word, "diversity", is used all the time when speaking about our environment, our ecosystems, where the change is going in the other direction. We are losing heirloom vegetables and crops, livestock, and fruit trees at an alarming rate as "big agriculture" promotes only the highest yield varieties of everything.

While this has been happening, homogenization of our communities is also happening. The homogenization of culture, to me, seems to be akin to the monoculture of farming and forestry. This makes me wonder if we are promoting diversity in the wrong way. Maybe, we aren't actually promoting it at all.

The same smattering of gas stations, fast food joints, big box stores, and convenience stores littered the landscape up and down the highways throughout the whole trip. The "Indian" reservation at Grand Portage, MN had a casino… which seemed to be about the only things going on there, and this seemed like a sad fit to me for a way for Native Americans…or anyone for that matter…to make a living.

The last place we visited was the Menominee Indian Tribe's forest in Wisconsin. This is the only place in the United States where a Native American tribe managed to keep continuous control of, and actively manage their ancestral forestlands from pre-settlement times all the way through to the present day. Their story is a sad one. The Federal government opposed them every step of the way, and they have had to fight through numerous lawsuits to maintain the right to manage their own land over their history.

The Menominee Indian Reservation provides a case study in what is possible when such Native American philosophy towards managing natural resources is practiced. I visited the Menominee Reservation and toured their forest management operations and sawmill again while attending the 2017 Society of American Foresters convention. The impressive forest looks like "old growth", and gives the impression that the forest has never been harvested. However, responsible forest management here has been supporting the tribe for over 150 years.

According to their oral history, the early chiefs directed the tribe in the following manner: "Start with the rising sun, and work toward the setting sun,

but take only the mature trees, the sick trees, and the trees that have fallen.
When you reach the end of the Reservation, turn and cut from the setting sun
to the rising sun and the trees will last forever."

We Americans are a diverse mix. This diversity has given us many wells of
knowledge and experience to draw from; including America's first peoples. Yet,
the "one size fits all" policy imposed upon us all is resulting in the poverty of
mono-cultures in our crops, our forests, and our communities. True diversity
includes encouraging a community where a diversity of ideas can freely be
expressed, experimented with, blended, and adopted organically…without
coercion or suppression.

Earth Day, April 22nd, was established in 1970 through the efforts of
Wisconsin Senator Gaylord Nelson. His April 22nd speech in Denver Colorado
announcing the first Earth Day includes the following excerpt:

"Our goal is a new American ethic that sets new standards for progress,
emphasizing human dignity and well-being rather than an endless parade of
technology that produces more gadgets, more waste, and more pollution."

Having been born in 1970, I must say that during my lifetime, this goal has
certainly not been reached. While progress has been made, it seemed to me that
during the '80's, Madonna's "Material Girl" song pretty well summed up the
prevailing psyche of America just a few years after the first Earth Day: we are
"living in a material world".

The Environmental Protection Agency was also created in 1970, and many
new laws and regulations have been introduced since that time to deal with air
and water pollution. Progress has been made in some ways; yet the goal of Earth
Day was for a new "ethic" to prevail in America. This ethic would focus on
quality of life rather than quantity of "stuff" in its measure of progress.

Ironically, one interesting example of someone who lives this Earth-friendly
ethic…at least in the area of his housing… is one of the world's richest men:
Warren Buffet. Worth over $50 billion, Mr. Buffet still resides in the Omaha
home that he purchased in 1958 for $31,500. He owns no plane or yacht, but
donates hundreds of millions of dollars to various charities and causes around
the world.

Warren Buffet's middle son, Howard Buffet, lives on a farm a few miles
from where I grew up in Illinois. He was often seen at farm auctions and local
events when I was growing up around there, and looks just like any other farmer
there. I was at an auction one time as a kid, and bid on a box full of various

kinds of steel foothold and conibear traps I hoped to buy for my friend and myself to use to trap muskrats from his pond. He noticed he was bidding against a kid and stopped bidding. Although he is worth millions, instead of flaunting his wealth and "living it up", he chooses to spend much of his time working to help improve conservation and farming; particularly in Africa. According to a Wall Street Journal article titled "A Buffet Turns to Farming in Africa", after taking a trip to Africa to photograph wild animals, Mr. Buffet realized that efforts to protect wildlife habitat in Africa would fail without first fighting its food shortage.

"I'm watching this thinking, 'They are going to destroy the last forest.'" Mr. Buffet later recalled. "It was an epiphany for me: The hungry can't worry about conservation. I realized you can't save the environment unless you give people a chance to feed themselves better." Much of Buffet's efforts in Africa include introducing higher yield, yet sustainable farming practices which conserve the soil and do not require the use of expensive fossil fuels, chemicals and equipment and so can be practiced in poor regions. http://www.thehowardgbuffettfoundation.org/

Yes, a new "land ethic" is what Earth Day is about. Without it, the marketplace will continue to give people what they want… an endless parade of technology and gadgets… while continuing to convert the Earth's priceless resources into trash instead of sustainable solutions for a clean environment.

Indian Fort Mountain (2010). The forest changes with the elevation as one hikes up the mountain…even on this knob that isn't by most people's standards a "real mountain", just 1,600 feet elevation. First are the toe slopes; with oak-hickory forest on fairly productive silt loam soils. The grade steepens and the side slopes become thin-soiled, with shale. The white oak, *Quercus alba*, largely gives way to chestnut oak, *Quercus montana*; while the coves created by the meandering terrain are populated mostly with yellow poplar. About half way up the slopes is a "bench" that runs around the contour of our forest at about 1,000 feet elevation and varies from about the width of a roadway to several hundred feet wide. The level bench receives moisture from the hillside above, and grows some the highest quality trees on our forest; often with more yellow poplar but also black walnut (*Juglans nigra*) sugar maple (*Acer saccharum*), more white oak, northern red oak, and chinquapin oak (*Quercus muehlenbergii*). On these fertile bench lands, whose soils are improved by the limestone cap above, the Native Americans once tilled the land and later settlers continued to farm it.

On the slopes above the bench, more chestnut oak usually prevails, and then on top of the mountain the species composition depends largely on whether the mountain has a sandstone "cap" or is limestone…as well as the thickness of the soil. Thin soils feature mainly stunted chestnut oak and Virginia pine, while the deeper soils feature a mix of all the species found below. Indian Fort Mountain has thick enough soil over the center portion of the sprawling formation such that it was formerly farmed. One may have difficulty now believing there was once a farm on top, but indeed these "mountain tops" often were farmed or were used for orchards…above the pooling early frosts that could freeze their developing fruits.

I read the landscape and the forest. The hiking trail mostly followed an old roadbed to the top, where the younger trees and "old field" appearance told me the area had been cleared at some point. Coming to Indian Fort Lookout, I saw the beautiful views and it reminded me of Garden of the Gods in Southern Illinois…one of my favorite places. Under *Indian Fort Lookout* is a rock house cave called *Devil's Kitchen*. It reminded me of Cave-in-Rock; also in Southern Illinois…another of my favorite places. Coming out of the cave and trekking back up to Indian Fort Lookout, I was not aware that this is where Silas Mason stood when he received his vision for the Berea College Forest. I was there the day before my job interview; contemplating my own vision for this beautiful forest if I was so fortunate to be hired as the College Forester.

The West Pinnacle is where I obtained my own vision for the Berea College Forest. I could see Berea and the College Campus from this vantage point. So close to town, so close to a college, this could become a community forest much like those I had only heard about and read about, in Germany. The idea portrayed in the name "Indian Fort" seemed so opposite of my vision. The word "Indian" perpetuated both the wrong use of the word "Indian" and an emphasis that the place was no longer theirs, but now "other than theirs". The word "Fort" implied, to me, that the place featured a structure known to have been a defensive bastion against attackers. No, this place needed to become a community forest where all were welcome. I returned to this spot after attending Berea College's "Civil Rights Tour" and with Martin Luther King's "Mountaintop Speech" freshly instilled in my mind, rededicated myself to this community forest ideal. This place, I promised myself, I would help to make a place which would resonate with folks as they looked over the treetops towards the Berea College campus and reflected on its history.

I had no idea there was no "Indian Fort" here as I wandered around looking for the stone "walls" that were supposed to be here somewhere. There wasn't anything resembling a fort anywhere to be found. "Did Indians even have forts?" I thought to myself as I contemplated the name. Well, it turns out there are some very low "walls" that occupy several locations around the perimeter of the natural flat-topped mesa-like geologic features that is Indian Fort Mountain. These linear rock piles are barely discernable to the untrained eye, but were supposed by early settlers to be placed this way by Native Americans. Imagining that the "Indians" used this "Masada-like" place as a fort, it was named "Indian Fort Mountain".

No one really knows why the stone "walls" were built, or who built them. Assuming it was Native Americans, they may have used the place for ceremonial uses, or maybe they just piled the rocks over along the edges of the mountain for some utilitarian purpose rather than defensive. Just below the cliff of Indian Fort Mountain; right below "Indian Fort Lookout", is a large rock house (what we call here a cave-like overhand under a cliff) that Native Americans undoubtedly used for shelter. Rock houses like this are pretty common along the cliffs in this area. Unfortunately, pillagers over the years have dug these rock houses in search of artifacts, arrowheads and the like.

I grew up watching Westerns, and often found myself rooting for the Indians. Their supposed way of life… whether living in the pueblos of the southwest, teepees of the Great Plains, or wigwams and long houses of the Eastern forest, is fascinating to me. America's westward expansion rapidly transformed one of the great wild areas of our planet. Immense herds of bison were almost completely extirpated; vast swaths of prairie put under plow; once wild mountain ranges subjected to mining and cattle grazing. The way of life for America's first inhabitants, the American "Indians", quickly changed forever. We would do well to learn what we can of the Native Americans and their ways.

The loss of the Native American way of life was deeply disturbing to George Catlin; a man who had the vision, the courage, and the artistic ability to document their way of life before it was too late. Recognizing that disaster was imminent, Catlin traveled west five times in the 1830s just after the United States Congress passed the Indian Removal Act. His mission was to paint American Indians and their way of life to "rescue from oblivion their primitive looks and customs".

Catlin, the first artist to record the Plains Indians in their own territories, admired these people and considered them an example of the En-lightenment ideal of "natural man" living in harmony with nature. More than 500 paintings were produced; many of which provide the only documentation of Native American customs and culture.

Catlin's paintings are quite famous today and are recognized as a real cultural treasure. However, during his lifetime, he received little support for his efforts. He lobbied the U.S. government for patronage funding many times, but was turned down. His paintings were well received in Europe, but he found that audiences were more interested in sensationalism, and he resorted to presenting real Indians acting out war dances to help ends meet in his travels. Even with these efforts, he went bankrupt in 1852.

Fortunately, Catlin's paintings survived to the present day through the philanthropy of a Philadelphia industrialist who paid Catlin's debts to acquire the paintings and then donated them to the Smithsonian Institute. To learn more about Catlin and to see his paintings, see: http://www.americanart.si.edu/catlin/highlights.html

Catlin's paintings depict Native Americans in such detail that much can be learned of their customs, art and clothing from his meticulously detailed works. However, besides documenting Native American people, there is also great value in what can be learned about the environment prior to European settlement of the West. The vast landscapes featured in the paintings often depict, for instance, vegetation and wildlife as it once existed. Some feature Indians setting fire to the prairie, hunting bison, and utilizing resources in a way that would be described as "sustainable" in today's lingo.

"Behold, my brothers, the spring has come; the earth has received the embraces of the sun and we shall soon see the results of that love! Every seed has awakened and so has all animal life. It is through this mysterious power that we too have our being and we therefore yield to our neighbors, even our animal neighbors, the same right as ourselves, to inhabit this land."

– Tatanka Yotanka (Chief Sitting Bull), Hunkpapa Lakota Sioux

---

SIU-C Campus, Carbondale, IL (1989). Sweetgum trees lined the sidewalks along one portion of wide sidewalks that linked the Agriculture building to the campus library where I worked for one semester. The prickly sweetgum balls would fall onto the sidewalk and I couldn't help but laugh at the barefoot group of supposed students who were gingerly navigating around them as I caught up to them and sought to pass by on my way to work. I suppose they targeted me because I was wearing cowboy boots and a cap…or maybe they saw I had come from the Agriculture building. For whatever reason, they started screaming at me as I approached. I couldn't even tell what they were saying, as they all screamed some chant over and over at the same time but not in sync.

This was my first experience with "tree huggers. The group walked around campus barefoot, went on hunger strikes, spiked trees designated for harvest on the Shawnee National Forest, and screamed at us forestry students when we tried to talk with them. I went with a couple other forestry students to one of their advertised meetings and we brought along some pamphlets we had made up to give them. The pamphlets explained how the particular harvest they were protesting was actually prescribed to remove the planted pine trees to allow the native hardwoods which had now grown up under them, to grow…it was for forest restoration. As we handed them out, every one of them was promptly hurled back to us wadded up and unread. We began to feel like we were going to get beat up if we stayed, so we got out of there.

So, where did the "tree hugger" movement popularized in the 1970's and active into the late 1980's and beyond get started? India. The Chipko movement became famous among environmentalist circles in the early 1970's when poor villager women protested the cutting of trees in the Uttarkhand region in the Himalayan foothills of India by joining hands around the trees. The government reportedly sent in troops to shoot them, but the troops refused to follow through. "Chipko" means "embrace" in their language. The success of this "tree hugging" spread throughout the region, and eventually lead to the passing of various national and state-level laws restricting the cutting of trees.

Rarely, however, is the whole story told.   The villagers were not so much against the cutting of the trees in the first place; but rather were protesting the trees being cut by outside contractors instead of them being permitted to cut them for themselves for local use. The new laws and regulations…the "Chipko laws"…actually resulted in a paralyzing, bureaucratic, inefficient and corrupt system that made it worse, not better, for the villagers to have access to timber and to build homes, roads and create much needed jobs…and an underground "timber mafia" came about as a result.

One of the "Chipko women" who was part of the movement in the first place, put it this way: "Now they tell me that because of Chipko the road cannot be built, because everything has become paryavaran [environment]... We cannot even get wood to build a house ... I plan to contest the panchayat [village council] elections and become the pradhan [mayor] next year ... My first fight will be for a road, let the environmentalists do what they will."

- Gayatri Devi, one of the "Chipko women" in Chamoli district.

(Shome 2008) https://thebreakthrough.org/archive/the_tree_huggers_of_india_the)

But even if the consequences of activism are sometimes perverse, the "tree huggers" have done something essential. Absent folks willing to take a brave and bold stand against the cutting of trees, we might only have photos remaining at this point of the great sequoia trees…the most massive living things on the planet. John Muir is perhaps the best known individual in American history in this regard. His tactics were brave and bold, but focused in an orderly fashion to educate and inspired others; not intimidate and frighten them.

Known for his activism, Muir helped save the Yosemite Valley, Sequoia National Park, and other wilderness areas. Without John Muir, there may not have been any giant sequoias and redwood trees left for our generation to be awed by. Muir's letters, essays, and books of his experiences and adventures in the Sierra Nevada Mountains of California introduced Americans to the idea of wilderness preservation. John Muir also founded the Sierra Club; one of the most important conservation organizations in the United States.

Muir was born in Dunbar, Scotland. His family immigrated to the United States when he was a young boy and settled in Wisconsin. He grew up in a very strict and religious household where he learned, by age 11, to recite much of both the New and Old Testament. At age 22, Muir enrolled at the University of Wisconsin at Madison. Here, he studied geology and botany for two years, but never graduated. Instead, he left school to wander the Canadian wilderness collecting plants. Muir eventually had to come out of the wilderness and he worked for a short time with his brother at a saw mill. Later, he worked as an industrial engineer at a carriage parts plant in Indiana. Here, he proved to be an ingenious inventor and might have stayed in this profession if it had not been for an accident at the plant when a tool he was using struck him in the eye.

Muir had to stay in a darkened room for six weeks; wondering if he would regain his sight. When he did, he had a renewed purpose. Muir would later write "This affliction has driven me to the sweet fields. God has to nearly kill us sometimes, to teach us lessons. ". From that point on, he followed his dream of exploration and the study of plants.

Muir took off on a 1,000 mile walk from Indiana to Florida, with no specific route chosen other than to follow the "wildest, leafiest, and least trodden way I could find". When he reached Florida, Muir hoped to board a ship to South America and continue his exploration of plants there. However, he contracted malaria while in Florida and had to abandon his plans. Instead, he decided to go to California. Once Muir arrived in California, the rest is history.

India may owe its best early forest protection efforts to another German forester. Most foresters have heard of Gifford Pinchot; the first Chief of the United States Forest Service and the 28th Governor of Pennsylvania. However, few have heard of his mentor and teacher, Dietrich Brandis. It was Brandis whom Pinchot turned to, to help him formulate a strategy to construct an effective government forest service. ("Forestry by Correspondence", Forest History Society.)

Brandis never set foot in America, but he advised Pinchot through personal letters which have been preserved by the Forest History Society. These letters detail how Brandis' success in bringing German forest management to India became Pinchot's model for the United States Forest Service. Dietrich Brandis was born in Bonn Germany in 1824, and earned a doctorate in botany from the university there in 1849. He taught for six years, and then was recruited for the position of Superintendent of Forests in the Pegu province of Burma. At the time, all of what is now India, Pakistan, Bangladesh and Burma were under the rule of the British Empire.

Brandis immediately introduced scientific forest management and sustainable forestry practices throughout Burma and India. Some of his work included the determination of peak volume, rate of growth, optimum rate of harvest, and forest protection plans against pests and fire. Where exploitation had formerly been the matter of course, Brandis established purchase and harvesting rules and set up management areas he called conservancies. These conservancies were then staffed with conservators to manage and protect them. This system became the model for our Forest Rangers in the United States.

After seven years, Brandis was given the title of Inspector General of Forests in India. He held this position for 20 years, developing forest legislation and helping to create research and training institutions; including the State Forest Service College at Dehra Dun. Pinchot mirrored these efforts in the United States, and helped to establish the Yale School of Forestry.

Perhaps one of the most important influences Brandis had on Gifford Pinchot was his warning of "dishonesty and corruption" of staff who would necessarily be working in faraway places. His solution was to implant an "esprit de corps" encouraging "free interchange of experiences and opinions upon professional matters" among foresters.   Pinchot responded by creating the Society of American Foresters.

After Brandis retired, he remained active in Indian forestry issues, and at the age of 75 he began to write his botanical work "Indian Trees"; which covered 4,400 species of tropical trees. Twelve species of trees and one genus are named after this great man who combined the practical and the literary to enhance forestry. Mr. Brandis passed away in 1907. Gifford Pinchot; honoring him in a memorial address, said:

"… no other person's achievements rank higher or had a similar influence on forestry during the latter half of the past, and at least on the beginning of this century." (Quote attributed to "Leben und Werk von Dieterich Brandis", by Herbert Hesmer, 1975.)

Mountains often exhibit a rich diversity of plant species due to the gradient in elevation that provides a changing matrix of soil type, moisture level, aspect, and shade. This combination of the mountainous terrain, as well as a changing climate over history that included glaciation, has made the Southern Appalachians one of the most biologically diverse temperate regions in the world. It seems fitting that Berea College would have been founded in this region…albeit on the edge of Appalachia and Kentucky's "bluegrass" region. Seemingly always "on the edge", the story of Berea College is one of diversity, bravery, and faith from the beginning.

From https://www.berea.edu/about/history/:
The startled railroad surveyor dropped his notebook as his surveying instrument

focused on a brick structure extending above the forest canopy.

Ladies Hall, Berea College's first brick building, seemed totally out of place in the woodland setting.

"Whoever put up that building in this wilderness must have had faith," the surveyor observed.

The surveyor's experience came some 20 years after the Rev. John G. Fee started a one-room school in 1855 that eventually would become Berea College. Fee, a native of Bracken County, Ky., was a scholar of strong moral character, dedication, determination and great faith. He believed in a school that would be an advocate of equality and excellence in education for men and women of all races.

Fee's uncompromising faith and courage in preaching against slavery attracted the attention of Cassius M. Clay, a well-to-do Kentucky landowner and prominent leader in the movement for gradual emancipation. Clay felt he had found in Fee an individual who would take a strong stand on slavery.

In 1853, Clay offered Fee a 10-acre homestead on the edge of the mountains if Fee would take up permanent residence there. Fee accepted and established an anti-slavery church with 13 members on a ridge they named "Berea" after the biblical town whose populace was open-minded and receptive to the gospel (Acts 17:10).

In 1854, Fee built his home upon the ridge. In 1855, a one-room school, which also served as a church on Sundays, was built on a lot contributed by a neighbor. Berea's first teachers were recruited from Oberlin College, an anti-slavery stronghold in Ohio. Fee saw his humble church-school as the beginning of a sister institution "which would be to Kentucky what Oberlin is to Ohio, anti-slavery, anti-caste, anti-rum, anti-sin." A few months later, Fee wrote in a letter, "we…eventually look to a college — giving an education to all colors, classes, cheap and thorough."

Fee worked with other community leaders to develop a constitution for the new school, which he and Principal J. A. R. Rogers insisted should ensure its interracial character. It also was agreed that the school would furnish work for as many students as possible, in order to help them pay their expenses and to dignify labor at a time when manual labor and slavery tended to be synonymous in the South.

The first articles of incorporation for Berea College were adopted in 1859. But that also was the year Fee and the Berea teachers were driven from Madison County by Southern pro-slavery sympathizers. Fee spent the Civil War years raising funds for the school; in 1865, he and his followers returned. A year later, the articles of incorporation were recorded at the county seat, and in 1869 the college department became a reality.

The first catalog, issued for 1866-67, used the corporate name "Berea College," but the title "Berea Literary Institute" was printed on the cover because it was thought to convey better "the present character of the school." Enrollment that academic year totaled 187 — 96 black students and 91 whites. For several decades following the Civil War, Berea's student body continued to be divided equally between white and black students, many of whom went on to teach in schools established solely for African-Americans.

In 1886-87, the school had three divisions: Primary, Intermediate and Academic. Students could pursue a college preparatory course, a shorter course, or a teachers' course. In 1869-70, five freshmen were admitted to the College Department, and in 1873 the first bachelor's degrees were granted.

Berea's commitment to interracial education was overturned in 1904 by the Kentucky Legislature's passage of the Day Law, which prohibited education of black and white students together. When the U.S. Supreme Court upheld the Day Law, Berea set aside funds to assist in the establishment of Lincoln Institute, a school located near Louisville, for black students. When the Day Law was amended in 1950 to allow integration above the high school level, Berea was the first college in Kentucky to reopen its doors to black students.

By 1911, the number of students seeking admission to Berea was so great that the trustees amended the College's constitution to specify the southern mountain region as Berea's special field of service. The commitment to Appalachia, however, began as early as 1858 when Rogers, after a trip through the mountains, identified the region as a "neglected part of the country" for which Berea was founded to serve.

Curricular offerings have varied at Berea to meet changing needs. In the early 1920s, in addition to its College Department, Berea had a high school that included ungraded classes for students who had not had educational opportunities, an elementary school, a vocational school and a Normal School for teacher training. Although the general mission of serving students with financial need continued, units and divisions were reorganized through the years. In 1968, Berea discontinued its elementary and secondary programs and now focuses entirely on undergraduate college education.

Berea's distinctive commitments and educational programs have brought the College national recognition.

Several years ago, I had an opportunity to accompany a group from Berea College on their "Civil Rights Tour". The tour was sponsored by the Carter G. Woodson Center at Berea College and made possible by a grant. During the tour, we visited sites important to the Civil Rights struggle in Birmingham, Montgomery, and Selma, Alabama, as well as Memphis, Tennessee. The tour was very educational and inspiring. It was also a lot of fun; with great camaraderie and some wonderful food.

I suppose it is fitting that Berea College would have such a tour, since it was the first interracial college in the South. Berea College's position that "God has made of one blood all peoples of the earth," is still central to the College's culture and programs. However, it is easy to fall into complacency, and to lose focus on the vision of a world shaped by these values, such as the power of love over hate, human dignity and equality, and peace with justice.

The Civil Rights Tour reinvigorated me spiritually; much like a revival. A key element was listening to Martin Luther King's speeches at the locations they were delivered in the 1960's. Most moving was his prophetic last speech… "I've been to the Mountaintop"… delivered April 3rd, 1968 in Memphis, TN, just before he was assassinated. I saw then how the forests and the mountains of Berea College are essential symbols for the College's mission of interracial education.

If you have never heard "the mountaintop speech", or it has been a while since you have, I would encourage you to search it out online and listen to it. Dr. King finishes the speech with the following words:

"Like anybody, I would like to live a long life. Longevity has its place. But I'm not concerned about that now. I just want to do God's will. And He's allowed me to go up to the mountain. And I've looked over. And I've seen the Promised Land. I may not get there with you. But I want you to know tonight, that we, as a people, will get to the Promised Land! And so I'm happy, tonight. I'm not worried about anything. I'm not fearing any man! Mine eyes have seen the glory of the coming of the Lord!"

When I look out over the West Pinnacle and see Berea College in the distance below, I now think of this speech. Surely, to many of the freed slaves who were some of Berea College's first students, Berea College was a kind of "promised land". Many of these early graduates went on to achieve great things… to "climb that mountain"… and become some of the first African American leaders in our nation.

To learn more about Berea College's interracial history, visit the Carter G. Woodson Center for Interracial Education. It is in the Alumni Building, and is open to the public. The Center was established in October of 2011 to "promote social and cultural change through the trans-formative power of education that recognizes the enhancing value of all peoples of the earth". http://www.berea.edu/cgwc/)

I noted earlier in this chapter the important connections between cultural and ecological diversity. People of all races, nationalities, and cultures, when hiking on the trails at Indian Fort Mountain, naturally celebrate the diversity they see in the forest ecosystem around them. I think this exposure helps us to realize that human diversity should also be celebrated and it is not something that we should seek to homogenize. Each species of organism has its own unique niche in the ecosystem that it fills. Besides, could there be anything more boring than hiking in a field of GMO (genetically modified organism) corn?

"Survival of the fittest"…we have all heard the term. Darwin's theory of evolution by natural selection was first formulated in his book "On the Origin of Species by Means of Natural Selection, or the Preservation of Favoured Races in the Struggle for Life" in 1859. "Races" refers to species in the book, but I suspect the use of the word "race" has something to do with its title being usually shown in abbreviated form as simply "On the Origin of Species". Sadly, this concept, "survival of the fittest," has been used to justify the "preservation of favoured races" in human races as well…and even the extermination of the "weaker" by the "fittest".

The 1927 Supreme Court ruling in the case Buck v. Bell, for instance, permitted the state of Virginia to sterilize an "imbecile"…a scientific term of the day…along with "idiots" and "morons". The Buck v. Bell case was even cited by the Nazi defense in the Nuremburg Trials. America's leadership in eugenics isn't something that was talked much about after World War II, but it was embraced by many of our nation's leaders. (Cohen, 2017)

Beyond my concern about the misuse of the "survival of the fittest" concept, there is another problem with it, it seems to me, at its core. My sense of ecology, with its myriad examples of symbiotic relationships and great diversity of species speaks to a greater role of cooperation between species than was understood at the time of Darwin. Is it time to replace "survival of the fittest" with "survival of the fitting-in-est?"

I can't help but wonder if, had Goethe's phenomenological approach to science prevailed, we would all be discussing evolution in these terms…survival of the fitting-in-est. Andrew Dickson White (1832 –1918) cofounder of Cornell University, considered Goethe's ideas on evolution "the most brilliant of all" after listing Erasmus Darwin (Charles Darwin's father) and others as important contributors to this (then new) subject.

For Goethe, the production of new knowledge is inseparable from a *Geschichte des Denkens und Begreifens* (a history of thinking and conceptualization). In other words; knowledge has to do with association; not only about separation. My grandpa's maxim: "Order is the highest virtue" is Goethean in that it equates Order...with a capital O…the ultimate Order…with the ultimate Virtue. Goethe sought, with science, to penetrate into the crucial underlying, sensorial-invisible archetype-pattern (Ur-phänomen) of all things…the inherent Order in them…to understand them holistically.

It seems to me that, as the concept of an ultimate Order began to be discarded in science…it is no wonder that an ultimate Virtue… would have to die with it. Ernst Lehrs described the exclusively rationalist approach to science as "one-eyed, color blind". Teaching in Waldorf schools in the early 1900's, he emphasized a Goethean approach to science in teaching and dedicated his life's work to finding meaningful principles for life in our materialistic technological society.

Can meaningful principles for life be found in science? I believe they can...
and will...but only as science begins to rediscover itself after tiring of its reduc-
tionist-only methodology. There are signs this is happening. In 1998, Henri
Bortoft wrote *The Wholeness of Nature: Goethe's Science of Conscious Participation
in Nature,* for instance, where he discusses the relevance and importance of
Goethe's approach to modern scientific thought. Similarly, biologist Brian
Goodwin (1931-2009) claims in his book *How the Leopard Changed Its Spots:
The Evolution of Complexity* that organisms are dynamic systems which are the
primary agents of creative evolutionary adaptation.

Perhaps, as modern scientists begin to holistically put back together their
reductionist models, Goethe's "Ur-phänomen" will reveal itself. Ralph Waldo
Emerson, in his book *Nature* used the concept of a transparent eyeball as a repre-
sentation of an eye that is absorbent rather than reflective. Taking in all that
nature has to offer, the idea of the transparent eyeball serves as a tool for the
observer to "become one" with nature. Then, I suppose, both beauty and order
will be in the eye of the beholder and we may all behold how we, too, fit in.
Max Planck said it well: "Science cannot solve the ultimate mystery of nature.
And that is because, in the last analysis, we ourselves are part of the mystery that
we are trying to solve."

If we consider ourselves part of the mystery of nature that we are trying to
solve, it makes sense to examine ourselves as much as nature. Doing so, we will
not want to see it "one eyed and color blind" as Ernst Lehrs put it, but with
stereoscopic vision in full color...with all its inherent order and beauty. Odin is
said to have plucked out one of his eyes to obtain wisdom. In some ways, when
we look at science, it seems we do the same thing. But, I think the ancients
knew something we are only beginning to understand...these two orders merge.

Order, after all...like our eyes...comes as a pair: Grouping Order and
Symmetry Order. Grouping Order can be understood as the order we see in
particular things that are grouped together. The elements are made up of
different groupings of atoms; these are grouped into gases, liquids and solids...
up through the various kinds of life forms grouped into species. From atoms to
planets, solar systems, and galaxies; there is Group Order.

Then, what is Symmetry Order? While Grouping Order may be thought of
in terms of multiplication....where a formula is replicated as in a fractal (think
Mandelbrot set) to produce a seemingly infinite number of possible renditions
of itself... Symmetry Order can be thought of as using the opposite mathematical concept: division. Working the opposite of multiplication, Symmetry Order
is the order that seeks to merge the grouped things all back together into One.

Physicist David Bohm identified this "dividing" or "enfolding" type of order
and called it Implicate Order. In his book *Wholeness and the Implicate Order*
Bohm writes:

This order is not to be understood solely in terms of a regular arrangement of objects (e.g., in rows) or as a regular arrangement of events (e.g. in a series). Rather, a total order is contained in some implicit sense, in each region of space and time. Now the word 'implicit' is based on the verb 'to implicate'. This means 'to fold inward' (as multiplication means 'folding many times'). So we may be led to explore the notion that in some sense each region contains a total structure 'enfolded' within it.

Goethe seems to have been on the same track with his *Ur- Phänomen*.

I think the most important theme to remember when tasked with managing a forest is to keep all the "ingredients" or "possibilities" intact so that we are not taking from future generations any potentialities of what we have inherited. These potentialities lie not only in the capacity of the land, but in ourselves. How we develop is tied to the land that sustains us.

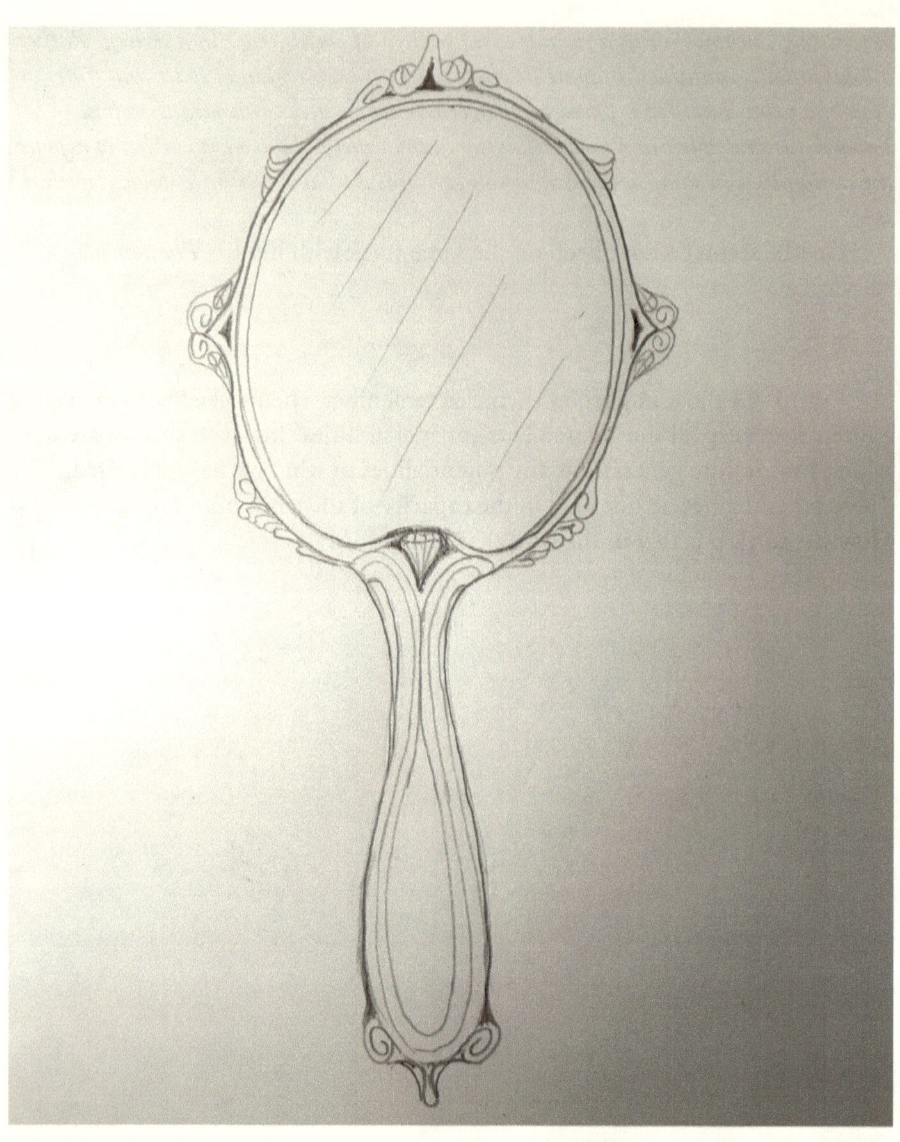

Chapter 4:
Know Thyself

"Education isn't something you can finish." – Isaac Asimov

I have been a member of the Society of American Foresters since 1990, and as a member, I have been receiving their publication, the Journal of Forestry, ever since. Mostly, the journal consists of scientific papers on forestry, generally very dull, statistics-laden, jargon-infused articles that put me to sleep. Typically, if I read them at all, I skip to the conclusions section.

It's not that I am not interested in science, or can't understand scientific papers. To the contrary, I have always had a high regard of science, inventions, discoveries, etc... I loved Star Trek and as a child I envisioned we would be colonizing other planets by now. However, as a first generation college graduate who grew up in a very rural area, it just all seems so unnecessarily "high falutin'" to me how matters of science are communicated. Also, it seems like science often ignores common sense, and can take the simplest things and make them overly complicated...creating a "mental bureaucracy" to wade through before making a decision.

The June, 2012 issue of the Journal of American Forestry, however, contains an article which I really enjoyed. It is titled "Listening and Learning from Traditional Knowledge and Western Science: A Dialogue on Contemporary Challenges of Forest Health and Wildfire". The gist of the article is that Native Americans used fire as a management tool on the landscape for thousands of years, and maybe now, after the "experts" have messed things up really bad, they should listen to the Indians.

Two ways of thinking and knowing are outlined in the article: Traditional Knowledge and Scientific Ecological Knowledge. Traditional knowledge is defined as "a body of culturally transmitted knowledge and beliefs about the relationships of living beings (including humans) with one another and with their environment". In contrast, Scientific Ecological Knowledge situates the "humans" as observers, conceptually separated from the environmental world, while "focusing on the control of nature and primarily concerned with theories of general interest and applicability". These two definitions fit my two types of Order characterizations.

Although scientific ecological knowledge has been extremely successful in manipulating the environment to maximize simple production and economic gain, many problems have been created that might have been avoided had traditional knowledge not been disregarded or relegated to an inferior status. As a consequence, both our environment and our human communities have suffered.

The article describes a more synergistic approach; describing how Federal agencies, researchers, university-educated Native Americans, and Native American elders are beginning to work together to formulate strategies to better manage forests using fire. These efforts are purposefully integrating traditional knowledge, rather than ignoring it, to find ways to better manage natural resources on tribal lands. All participants expressed that traditional knowledge combined with scientific knowledge, could provide a better framework for solving problems than either system alone.

This effort is particularly urgent, as traditional knowledge… normally passed down through showing, telling, and doing…rather than publication… is quickly disappearing from cultures everywhere. Mass media, public education, and the disintegration of the family unit are homogenizing the world's people, and discarding thousands of years of place-based knowledge and tradition. To my way of thinking, my idea for this book…that practicing forestry is an art… goes along with this concept. Practicing anything as an art requires visualization and creativity. Thus, I think practicing forestry as an art happens along the same lines, for example, of cutting hair. A barber or hair stylist doesn't start measuring and counting the hairs on their client's head when they sit down…they envision what the desired outcome is for the person's hair and go to work on it to create it as they conceive it.

I have always been a big fan of Native America. I have some Native American ancestry in my heritage, and am proud of that. But, traditional knowledge is not limited to any particular culture. There is place-based and experience-based knowledge all over…Berea's identity is most closely tied with African-American and Appalachian traditional knowledge, experience, and art. Those who strive to preserve and pass on traditions, knowledge, and the love of the arts and crafts… whether weaving, gardening, carving, storytelling, music, or what not, have an important task as well. I would make the case that forestry is a traditional art in Appalachia, too. Families that lived on the land years ago, for generations, learned to sustain the forested areas as a key resource in their subsistence livelihood.

Who knows where science will take us… perhaps we'll turn ourselves into bio-engineered "Borgs" like on Star Trek, where we can have all our techno-gadgets incorporated into our bodies. Maybe we'll destroy ourselves and the planet before then…or create artificial intelligence systems that will destroy us. What Man can do with knowledge is amazing, but it can also be scary. Ethics, beliefs, and faith supposedly have no place in science; but without them, where is our humanity? Einstein, perhaps the greatest scientist of the 20th century and often thought to be an atheist, expressed the need for both dimensions when he famously said, "Science without Religion is lame; Religion without Science is blind." I agree that Man needs more than "book learning"; more than science.

Education which focuses only on science does not "feed" the whole person. I would even say that as education and participation in the arts and spiritual instruction are abandoned in our society, this leaves a void in our thought life which gives rise to toxic thinking. Fear and despair seem to permeate our technical, media-drenched culture. Even young children are getting caught up in it…worrying about matters that were not even discussed or in some cases imagined just a generation or two ago.

Appalachian arts and crafts are still part of the student work program at Berea College. While this may seem like a quaint and nostalgic endeavor to some, and it produces only a little revenue through our College Crafts Department, I think the most important reason for this activity to continue is for mental health. Involving students with our draft horses…caring for them and slowly learning to work with them, has the same effect. Important skills are learned, and the culture of arts and crafts and horsemanship are preserved…but so is one's physical and, most importantly, mental wellness. I have helped with horse therapy and seen physically and mentally handicapped individuals, as well as soldiers with post-traumatic stress disorder, helped greatly by the calming effect of interacting with animals. Others find a similar result from doing arts, crafts, or exercising.

Appalachia is a place of great scenic beauty, but it also has a long history of exploitation, poverty, conflict, addiction, and poor health. The majority of our students come from Appalachia and many have struggled with these issues. What is science, the health care industry, politics, and the education system's solution? Prescription drugs?

When one of my children was getting on the bus to go to Washington, D.C. with their eighth grade class here at Berea, the school nurse was standing beside the bus collecting their prescription drugs…with a laundry basket. By the time the students were all on the bus, the laundry basket looked pretty full. I couldn't believe it. Something is seriously wrong with that. The next year, they went on a field trip with their school to the county morgue. They learned, from the coroner, that drug overdose is the leading cause of death in their home county.

What are people thinking? Surely, there is a better way to combat what is happening in our culture…not only in Appalachia, but increasingly all over.… with the depression, despair, and drug (both legal and illicit) drug addiction. The "Greatest Generation", as it is sometimes called, which grew up in the Great Depression and fought World War II, had it far rougher than today's generation. This was my grandparents' generation. Yet, they didn't seem to fall apart. What's different?

It seems to me that the "Greatest Generation", on the whole, got more exercise and ate food that was less chemical-laden and was perhaps healthier. This may have had some positive effect on their physical health compared to the last couple of generations. But, this cannot account for their seeming ability to handle hardship, danger, and the rapid changes in technology and other difficulties their generation experienced "in spades". I believe they handled problems better because of the difference in their thought life.

According to Dr. Caroline Leaf's book Who Switched Off My Brain: Controlling Toxic Thoughts and Emotions: "Research shows that around 87% of illnesses can be attributed to our thought life, and approximately 13% to diet, genetics and environment. Studies conclusively link more chronic diseases (also known as lifestyle diseases) to an epidemic of toxic emotions in our culture."

Today's people are not only exposed to "an epidemic of toxic emotions in our culture", but have fewer opportunities to get rid of these thoughts through positive mental-physical interactions with others. Community, and the sense of belonging and purpose that comes with it, cannot be replaced with drugs.

I was always artistic, as a child…and often was scolded in class for drawing instead of paying attention. My unconscious habit of humming to myself as I drew… once I really got caught up in my imagination… caused distraction. So, the teachers had to get after me and the other students made fun of me. Eventually, however, it helped me to get positive attention as students and teachers alike complimented me for my artwork and sometimes asked me to do drawings for them.   Especially in grade school, I could barely tolerate sitting in class unless I escaped into my imagination. There were windows all along the outside wall of every classroom. I often watched squirrels outside and imagined being a squirrel to get through the day. . This was before ADHD became a "thing" usually to be treated with drugs rather than allowing for behavioral differences in classrooms.

Last year, my wife and I drove by a public school near Rogersville, TN. We just drove through to see the historic buildings, as it is one of Tennessee's oldest towns and we wanted to see the home of Tennessee's first newspaper and post office. The historic part of the town was beautiful, but the sight of the nearby school made us get a sick feeling. As we approached it, I assumed it was a prison. Growing up in Illinois, I was used to seeing a lot of prisons. But, as we passed the windowless, dark-colored behemoth of a building, the sign read "Cherokee High School". What!? I turned around and pulled up to the sign as I couldn't believe the irony…a windowless, prison-like school was built among these beautiful mountains and then named after the Cherokee?

"No one there can look out the window and daydream at that place", I thought to myself as I tried to shake off a sunken feeling. Certain things bring back memories. I was interested in reading and writing, science and history, but struggled with math…one important memory for me is when I finally had a "Eureka moment" to understand multiplication. I am pretty sure I was the last in my class to comprehend it. No matter how much the teacher drilled it into us, all I could do was make guesses when it came my turn to blurt out a multiplication answer. Then, a young substitute teacher came to class one day. Right away, she understood I was guessing when I said the wrong answer to an easy multiplication question… "What is 4 X 5?" I believe. She drew five cookies on the chalkboard, then drew four chocolate chips on each cookie. Then, she said "These are five cookies. There are four chocolate chips, five times. So, 4 chocolate chips, times 5 cookies, is how many chocolate chips?" Immediately, I understood multiplication. People who learn like me, get it when math becomes concrete and visualizable.

I don't remember how the other students reacted, as I was mesmerized by the teacher and thinking about how easy this was now that I could visualize what it was. Sometimes, people will ask…derisively or jokingly: "Do I need to draw you a picture?" in response to someone not understanding them. Often the answer should be just, "Yes." Another saying goes "A picture is worth a thousand words", and I think pictures…art…is just as useful in teaching as it is for enjoyment.

I was excited to eventually discover that the great Greek and Roman artists that created such magnificent works as I was introduced to at the St. Louis Art Museum, were also inventors, mathematicians, and philosophers! Discovering Leonardo da Vinci was like discovering a real life super hero to me. The realization that the great artists of antiquity were often great mathematicians, philosophers, astronomers, and scientists, was a revelation to me.

My high school math teacher, "Mr. Webb", incorporated personal stories, science and history into his math classes. He was also the scholastic bowl coach. I was on the scholastic bowl team since the seventh grade. We were undefeated for a long time, and got to go to the National Scholastic Bowl in New Orleans my senior year. I specialized in history and geography, and practiced by drawing maps of the world…freehand, by memory…with all the nations, capitals, major rivers, mountain ranges, and other features. Each person on the team had a specialty…our team captain was such a whiz with literature and music that he could name practically any classical work and composer by hearing just a couple notes of the song. He not only knew the authors of classic literature, but had read nearly all their works. Our math specialist ended up working for NASA right out of college, and still does.

For me, math past arithmetic was so abhorrent that if not for Mr. Webb's way of teaching it, through stories and his witty and inspiring personality, I likely would not have attempted trigonometry or physics. Before he introduced math concepts, he would throw a paper wad across the room, behind his back, and into the trash can, flawlessly every time…and then start telling a story about Pythagoras, Euclid, Aristotle, or Socrates. He introduced math concepts through stories of their original conception by the ancient Greeks, and others.

His stories drew me in, as they touched on the purposes for the math discoveries; fighting wars and the inventions such as the screw, wedge and other early devices. Also political struggles, the topic of "the gods", and early advances in democracy were backdrops for formulas. Socrates, who paid for his open-minded education of the youth…considered to be "corrupting the youth" by the authorities…with his life… was the hero of the class and Mr. Webb had a bust of him on his desk.

So, I made good grades in physics and trigonometry…A's, even! I also developed an appreciation for how the Greeks made no distinction between philosophy and science. Great thinkers of those times might be experts in many fields; yet "fields" were not even really defined…knowledge was knowledge… and great thinkers might be accomplished artists, astronomers, playwrights, mathematicians and inventors all at the same time. They discussed such topics as "what is beauty?"; "what is justice?" in the same conversation as "what is the nature of matter?"

Mr. Webb made me a fan of Socrates, Plato, Aristotle, Euclid and other ancient Greeks…as well as Sir Francis Bacon and Sir Isaac Newton. So, when my grandpa introduced me to Goethe, I was primed to accept it and eager to learn from another great thinker and doer… in both art AND science. Great minds of the past spoke not only of things material, but things eternal… looking both upward and downward… toward the infinitesimal as well as the infinite…in whatever time or culture they lived in.

Where are these great minds today, in our world of technology, where the arts are barely taught in our education system? Such innovative teaching styles as Mr. Webb used would likely not be allowed in today's "teach to the test" straightjacket, while our professions increasingly separate between the arts and the sciences. He hired some of us on weekends to help him bale hay on his farm, dig post holes and whatnot. During breaks at school, he would play basketball with the jocks…always effortlessly winning with his famous trick shots. After scholar-bowl games, he'd take us to a pool hall if there was one near where our game was played, to reward us for winning…which was usually the case. There, he'd challenge us to games of pool where he shot only behind his back, or could only make bank shots…and still he always won. During study hall, he left his door open for anyone to come in and play a game of chess…he nearly always won.

The school I attended in St. Elmo, Illinois, has the large letters "Gymnasium" above the entranceway to the "gym" where basketball was played. The word gymnasium was first used in ancient Greece to describe a building used for both academic and physical education. The year I graduated (1988) our boys basketball team, the Eagles, had an exciting run through the state tournament and received fourth place…a big achievement for our tiny town. I was not on the basketball team…but the scholastic bowl team I was on did great that year too. We had the best record in the state and made it to the National tournament in New Orleans…though that generated far less excitement.

The popular song *St. Elmo's Fire (Man in Motion)*, by John Parr, served as the perfect theme song for our boys basketball team. The mind and the body are integrally connected and, like the Greeks, we should be educating the whole person to be a "Man in Motion" with something burning inside of them. Because:

Growin' up, you don't see the writing on the wall
Passin' by, movin' straight ahead, you knew it all
But maybe sometime if you feel the pain
You'll find you're all alone, everything has changed

Play the game, you know you can't quit until it's won
Soldier on, only you can do what must be done
You know in some way you're a lot like me
You're just a prisoner and you're tryin' to break free

I can see a new horizon underneath the blazin' sky
I'll be where the eagle's flying higher and higher
Gonna be your man in motion, all I need is a pair of wheels
Take me where my future's lyin', St. Elmo's Fire

Burning up, don't know just how far that I can go (just how far I go)
Soon be home, only just a few miles down the road
I can make it, I know I can
You broke the boy in me, but you won't break the man

I can see a new horizon underneath the blazin' sky
I'll be where the eagle's flying higher and higher
Gonna be your man in motion, all I need is a pair of wheels
Take me where my future's lyin', St. Elmo's Fire

I can climb the highest mountain, cross the wildest sea
I can feel St. Elmo's Fire burnin' in me, burnin' in me
Just once in his life a man has his time, and my time is now, I'm coming alive

I can hear the music playin', I can see the banners fly
Feel like you're back again, and hope ridin' high
Gonna be your man in motion, all I need is a pair of wheels
Take me where my future's lyin', St. Elmo's Fire

I can see a new horizon underneath the blazin' sky
I'll be where the eagle's flying higher and higher
Gonna be your man in motion, all I need is a pair of wheels

I can climb the highest mountain, cross the wildest sea
I can feel St. Elmo's fire burnin' in me
Burnin' burnin in me, I can feel it burnin'
Ohh, burnin' in side of me

"All things were in chaos when Mind arose and made order."
-ancient Greek saying attributed to Anaxagoras.

"The Greek Way", a book by Edith Hamilton, copyright 1938, has been one of the most influential books to me that I have ever read. I picked it up for free in the "to be disposed of" bin at a library while I was in college. I hope and pray that we, as a civilization, do not discard what the Greek's have given us: reason, freedom…Western civilization.

With the Greeks, something new came into this world. The Greeks were the first intellectuals in a world where the irrational had reigned for time immemorial…a world where enthroned despots imposed their whims and passions upon a wretched, subjugated populace…a world where a priestly organization sits astride the domain of the intellect like a lion across the neck of an antelope. The modern spirit, the "spirit of the West" as she calls it, is a Greek discovery. It is a gift to the world that we would do well to embrace more fully again. Despotism is always a threat to it, to rise again to squash it as the Persian Empire nearly did.

In the Persians of Aeschylus, a play written to celebrate the defeat at Salamis of the Persians, the Persian queen is told that the Greeks fight as free men to defend what is precious to them. She asks "Have they no master?" and is told no…that no man calls Greeks slaves or vassals. In the account by Herodotus, he adds: "They obey only the law."

The concept of the importance of the individual, and individual liberty… where the individual defends the state on his own free will, was a radical and new concept. "A slave is he who cannot speak his own thought", said Euripides.

A key concept to this freedom is for the individual to be educated. To teach the people so that they could think for themselves, would be to risk destroying the basis of a despot's power. The very meaning of the word philosophy… invented by the Greeks… is "love of knowledge". Indeed Plato, on writing of the differences between Greece and other countries, wrote "Egypt and Phoenicia love money. The special characteristic of our part of the world is the love of knowledge."

St. Luke commented that "The Athenians and the strangers sojourning there spend their time in nothing else but to tell or hear some new thing." Likewise, St. Paul, when travelling through Asia Minor, was mobbed, imprisoned and beaten everywhere he went until he reached Greece. Yet, in Athens, "they brought him unto the Areopagus, saying 'May we know what this new teaching is?'"

Only in Greece did the shrines, such as at Delphi…described by Plato…say such things as "Know thyself" and "Nothing to excess". What I like most about the Greeks is that they were not only great scientists and mathematicians, but also artists, sculptors, poets…and dreamers. They held no "vital lies". Accordingly, model scientist for his time, Aristotle, wrote:

"Since then reason is divine in comparison with man's whole nature, the life according to reason must be divine in comparison with (usual) human life. Nor ought we pay regard to those who exhort us that as men we ought to think human things and keep our eyes upon mortality: nay, as far as may be, we should endeavor to rise to that which is immortal, and live in conformity with that which is best, in us. Now, what is characteristic of any nature is that which is best for it and gives most joy. Such to man is the life according to reason, since it is this that makes him man."

The Greek Way, thus, was one of an enlightened intellect and spirit…not just one or the other…one at the expense of the other. Proof of this lies in their art…their genius bore fruit in both the sciences and the arts. This allowed for an extraordinary flowering of both the intellect and the spirit…they did not view the conclusions reached by the spirit as those reached by the mind as being opposed to one another. Truth is truth…the truth of poetry and the arts and the truth of science are both true. Mind and spirit are on equal footing. And so, love of reason and of life…delight in using both the mind and the body…where great minds produced advances in science, the outflowing of the spirit gave us magnificent works of art, and sculptured athletic bodies competed in the Olympics…are distinguishing features of the "Greek Way".

The Greek Way nearly was extinguished by the Persian king, Xerxes, when he invaded with his vastly larger army. The Persians were so numerous that their great clouds of arrows were said to hide the sky; upon which it is said Greek soldiers said "Good, then we will fight in the shade". Facing certain death, Leonidas the Spartan and his 300 men held off the invading army in the narrow pass called Thermopylae…while on the other side the Persian officers reportedly flogged their men to make them advance. The Greek leaders told their men "When we join battle with the Persians, before all else remember freedom."

After the astonishing victory over the vastly superior numbers of the invading Persians, it was thought that divine justice had secured their freedom. Yet, Greece was not perfect…fighting between Athens and Sparta resumed and internal power struggles emerged…the democracy became corrupt and full of vices and greed. The downfall of Greece may have been described best by Thucydides III: "The cause of all these evils was the desire for power which greed and ambition inspire". Edith Hamilton, in her book "The Greek Way", in which this section is based, put it this way:

"Freedom strictly limited by self-control—that was the idea of Athens at her greatest. Her artists embodied it; her democracy did not. Athenian art and Athenian thought survived the test of time. Athenian democracy became imperial and failed." While America could do well to learn from Athens with regard to preserving democracy, Edith Hamilton goes further. Even with the Renaissance and rediscovery of Greece, she claims we still have not recovered the balance between mind and spirit that exemplified ancient Greece. Rome applied this balance in the area of law, and became the lawgiver of the world, but in no other field has this happened since:

"People began to enjoy themselves and they were using their minds. They demanded liberty to think and to love life and the beauty of earth, but in their turn they ended by regarding as negligible the things that are not seen and they made their gain finally at the cost of morality and ethics. The Reformation asserted both morality and man's right to think for itself, but denied beauty and the right of enjoyment. The last great swing of the pendulum was in the nineteenth century when the battle was fought for scientific truth, and in the victory religion and art and the claims of the spirit were all slighted or discarded."

Four hundred years before Christ, Socrates comforted the world with his teachings that "goodness has a most real and actual existence", and that men can find it and help work it out. Aristotle, through Plato (a pupil of Socrates), wrote after Socrates died:

"There is a life which is higher than the measure of humanity: men will live it not by virtue of their humanity, but by virtue of something in them that is divine. We ought not to listen to those who exhort a man to keep to a man's thoughts, but to live according to the highest thing that is in him, for small thought it be, in power and worth it is far above the rest."

When I first read these words, I thought of the "still small voice" God used to speak to the prophet Elijah when he ran off to hide in a cave from Jezebel. Having read already the "Allegory of the Cave" account of Socrates conversation, recorded by Plato, I was astounded at the parallels between it and the Gospel. Some people believe that Socrates predicted the coming of the Christ. The movie "The Matrix" is based on this allegory…it illustrates that we are prisoners kept trapped in a cave by our ignorance…only able to see shadows that are mere vestiges of the true reality until someone can lead us out into the light…the journey upwards is the ascent of the soul into the intellectual world." Socrates predicts what would happen to whoever is able to lead us to the light. Glaucon, Plato's older brother and a philosopher himself, agrees.

SOCRATES: And if they can get hold of this person who takes it in hand to free them from their chains and to lead them up, and if they could kill him, will they not actually kill him?

GLAUCON: They certainly will.

––––––––––

"Order is the highest virtue", I remember my Grandpa would say, "Without it, we can't have anything else." Nature is orderly. Science is discovering this order in a greater detail than we have ever been able to imagine before…whether we look into the universe through a microscope or a telescope…the astonishing level of this order is mind-boggling. Interpreting how this order fits together is another matter…and especially asking what it means. Through the ages, great thinkers have contemplated such things.

Taken from The History Guide, by Kreis, 2000:

Around 600 B.C., Milesian thinkers "discovered" speculation after asking a simple but profound question: "what exists?" It was the Ionian natural philosopher, Thales of Miletus (c.624-548 B.C.), who answered that everything in the universe was made of water and resolves itself into water. What was so revolutionary about Thales was that he omitted the gods from his account of the origins of nature.

Anaximander of Miletus (c.611-c.547 B.C.), another Milesian thinker, rejected Thales, and argued instead that an indefinite substance—the Boundless—was the source of all things. According to Anaximander, the cold and wet condensed to form the earth while the hot and dry formed the moon, sun and stars. The heat from the fire in the skies dried the earth and shrank the seas.

Thales and Anaximander were "matter" philosophers—they believed that everything had its origin in a material substance. Pythagoras of Samos (c.580-507 B.C.) did not find that nature of things in material substances but in mathematical relationships. The Pythagoreans discovered that the intervals in the musical scale could be expressed mathematically and that this principle could be extended to the universe. In other words, the universe contained an inherent mathematical order.

Parmenides of Elea (c.515-450 B.C.), also challenged the fundamental views of the Ionian philosophers that all things emerged from one substance. What Parmenides did was to apply logic to the arguments of the Pythagoreans, thus setting the groundwork of formal logic. He argued that reality is one, eternal and unchanging. We "know" reality not by the senses, which are capable of deception, but through the human mind, not through experience, but through reason. This concept was to become central to the philosophic thought of Plato.

Perhaps the most important of all the Pre-Socratic philosophers was Heraclitus of Ephesus (fl. 500 B.C.). Known as "the weeping philosopher" because of his pessimistic view of human nature and "the dark one" because of the mystical obscurity of his thought, Heraclitus wrote On Nature, fragments of which we still possess. Whereas the Pythagoreans had emphasized harmony, Heraclitus suggested that life was maintained by a tension of opposites, fighting a continuous battle in which neither side could win a final victory. Movement and the flux of change were unceasing for individuals, but the structure of the cosmos constant. This law of individual flux within a permanent universal framework was guaranteed by the Logos, an intelligent governing principle materially embodied as fire, and identified with soul or life.

Fire is the primordial element out of which all else has arisen—change (becoming) is the first principle of the universe. Cratylus, a follower of Heraclitus, once made the remark that "You cannot step twice into the same river." The water will be different water the second time, and if we call the river the same, it is because we see its reality in its form. The logical conclusion of this is the opposite of flux, that is, a belief in an absolute, unchanging reality of which the world of change and movement is only a quasi-existing phantom, phenomenal, not real.

Democritus of Abdera (c.460-370 B.C.) argued that knowledge was derived through sense perception -- the senses illustrate to us that change does occur in nature. However, Democritus also retained Parmenides' confidence in human reason. His universe consisted of empty space and an infinite number of atoms (a-tomos, the "uncuttable"). Eternal and indivisible, these atoms moved in the void of space. An atomic theory to the core, Democritus saw all matter constructed of atoms which accounted for all change in the natural world.

What the Pre-Socratic thinkers from Thales to Democritus had done was nothing less than amazing—they had given to nature a rational and non-mythical foundation. This new approach allowed a critical analysis of theories, whereas mythical explanations relied on blind faith alone. Such a spirit even found its way into medicine, where the Greek physician Hippocrates of Cos (c.460-c.377 B.C.) was able to distinguish between magic and medicine. Physicians observed ill patients, classified symptoms and then made predictions about the course of a disease.

In todays' world, we have accomplished splitting the un-uncuttable atom, but do we really understand the universe any better than these ancient philosophers? It seems to me that the more we discover, the more wondrous the universe is seen to be. It bothers me that discussing the origins and meaning of the universe is discouraged in academia as well as in religious institutions.

Even in ancient Greece, questioning the prevailing status quo could get you in trouble. In 399 B.C., Socrates was charged by a jury of five hundred of his fellow citizens. His famous student, Plato, tells that he was charged "as an evil-doer and curious person, searching into things under the earth and above the heavens; and making the worse appear the better cause, and teaching all this to others."

Socrates was convicted to death by a margin of only six votes. Oddly, the jury then offered Socrates the chance to pay a small fine for his impiety. Yet, he rejected it! He also rejected the pleas of Plato and other students who had a boat waiting for him to help him escape. But Socrates refused to break the law. If he broke the law, he would be no citizen at all…and after visiting with his friends, he drank the fatal dose of hemlock. The parallels between Socrates and Jesus are quite incredible to me. The Mind that made chaos into order would come down and show men the light…lead them out of their cave…but he would be killed too, by men who hated the light.

The charge made against Socrates -- disbelief in the state's gods – were thought to corrupt the young if preached publicly. Meletus, the citizen who brought the indictment, sought precedents in the impiety trials of Pericles' friends. Incredibly, the Delphic oracle is said to have told Chaerephon that no man was wiser than Socrates…and during his trial Socrates even used this as justification…claiming that in exposing falsehoods, he had proved the god right!

He at least knew that he knew nothing. Socrates' irony is funny, but he clearly believed that his mission was divinely inspired.

Socrates did not reveal answers or truth; he just questioned everything. What is courage? What is virtue? What is duty? What is beauty? What is justice? So, he taught his students to discover, to think…and they realized most people could not answer these fundamental questions…yet all of them claimed to be courageous, virtuous and dutiful. (http://www.historyguide.org/ancient/lecture8b.html)

Nikola Tesla…perhaps the greatest inventor and thinker of our modern times…claimed: "My brain is only a receiver, in the Universe there is a core from which we obtain knowledge, strength and inspiration. I have not penetrated into the secrets of this core, but I know that it exists." He advised: "If you want to find the secrets of the universe, think in terms of energy, frequency and vibration", and predicted: "The day science begins to study non-physical phenomena, it will make more progress in one decade than in all the previous centuries of its existence." Tesla, by the way, was a big fan of Goethe and reportedly could quote Faust from memory in several languages. He was also good friends with Mark Twain…of all people. Hannibal, Missouri, was another one of my favorite places to visit growing up and he was one of my favorite authors. I'd imagine floating down the Mississippi on a raft and exploring the river towns and wooded banks on the way down to New Orleans.

In quantum physics, science is considering energy, frequency and vibration…and finding interesting things out that seem about as strange as some of Tesla's ideas. Ever heard of the "double slit experiment"?

An article in Popular Mechanics, 2016, explains it this way:

Basically, waves that pass through two narrow, parallel slits will form an interference pattern on a screen. This is true for all waves, whether they're light waves, water waves, or sound waves.

But light isn't just a wave, it's also a particle called a photon. So what happens if you shoot a single photon at the double slits? Turns out, even though there's only one photon, it still forms an interference pattern. It's as if the photon travels through both slits simultaneously.

But wait, because it gets even weirder. As a new episode of PBS's Space Time shows, just by observing the double-slit experiment, the behavior of the photons changes.

The idea behind the double-slit experiment is that even if the photons are sent through the slits one at a time, there's still a wave present to produce the interference pattern. The wave is a wave of probability, because the experiment is set up so that the scientists don't know which of the two slits any individual photon will pass through.

But if they try to find out by setting up detectors in front of each slit to determine which slit the photon really goes through, the interference pattern doesn't show up at all. This is true even if they try setting up the detectors behind the slits. No matter what the scientists do, if they try anything to observe the photons, the interference pattern fails to emerge.

It gets even weirder than that.

A group of scientists tried a variation on the double slit experiment, called the delayed choice experiment. The scientists placed a special crystal at each slit. The crystal splits any incoming photons into a pair of identical photons. One photon from this pair should go on to create the standard interference pattern, while the other travels to a detector. Perhaps with this setup, physicists might successfully find a way to observe the logic-defying behavior of photons.

But it still doesn't work. And here's the really weird part: It doesn't work regardless of when that detection happens. Even if the second photon is detected after the first photon hits the screen, it still ruins the interference pattern. This means that observing a photon can change events that have already happened.

https://www.popularmechanics.com/science/a22280/double-slit-experiment-even-weirder/

Equally "weird" or "mind boggling" is the concept of infinity and of nothing. The idea that "everything came from nothing"…whether in a "big bang" or otherwise…is difficult to fathom or explain. Does "nothing" still exist…or, if not, will it again? First, you have to define "nothing", and this gets interesting. Charles Seife, in his book Zero; The Biography of a Dangerous Idea: "Zero is powerful because it is infinity's twin. They are equal and opposite, yin and yang. They are equally paradoxical and troubling. The biggest questions in science and religion are about nothingness and eternity, the void and the infinite, zero and infinity. The clashes over zero were the battles that shook the foundations of philosophy, of science, of mathematics, and of religion. Underneath every revolution lay a zero – and an infinity."

So, can zero be both nothing and everything simultaneously? Is that how quantum physics works? According to the Gevin Giorbran (everythingforever.com) "What zero is not, is nonexistence. As Parmenides said long ago, nonexistence cannot be. There is no state 'greater than' or 'less than' the perfect zero. Zero is the default setting of reality."

Berea College is, of course, an educational institution and as such is in the business (as a non-profit) of education. Berea College…like the ancient Greeks…has not been concerned with only the mind, but also the spirit. Berea College's Preamble and Eight Great Commitments (taken directly from Berea College website).

Berea College, Founded by ardent abolitionists and radical reformers, continues today as an educational institution still firmly rooted in its historic purpose "to promote the cause of Christ." Adherence to the College's scriptural foundation, "God has made of one blood all peoples of the earth (Acts 17:26)," shapes the College's culture and programs so that students and staff alike can work toward both personal goals and a vision of world shaped by Christian values, such as the power of love over hate, human dignity and equality, and peace with justice.

Only the first part of Acts 17:26 is paraphrased. I am quite sure the founders of Berea College knew the entirety of it. Let's read the full sentence where this verse is found, where Paul addressed the crowd at Mars Hill in Athens:

"God that made the world and all things therein, seeing that He is Lord of heaven and earth, dwelleth not in temples made with hands; Neither is worshipped with men's hands, as though he needed any thing, seeing He giveth to all life, and breath, and all things; And hath made of one blood all nations of men for to dwell on all the face of the earth, and hath determined the times before appointed, and the bounds of their habitation; That they should seek the Lord, if haply they might feel after him, and find him, though He be not far from every one of us: For in him we live, and move, and have our being; as certain also of your own poets have said, For we are also His offspring." (Acts 17:24-28)

To "know one's self" we would do well not only to focus on matters of the body and of the mind, but also of the spirit. Let's go on a hike, on a walkabout, on a spiritual journey.

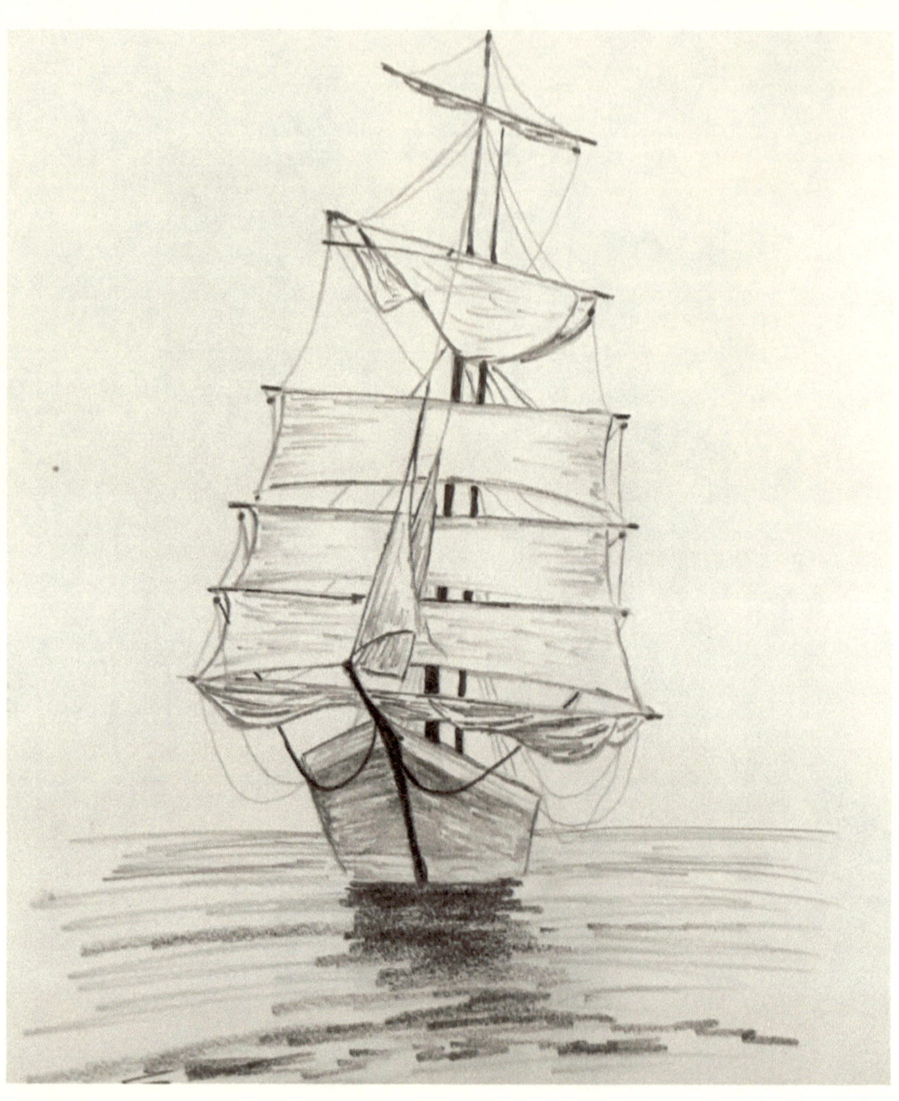

Chapter 5:
Journeys

Hickory Creek, IL (around 1979) "Honk! Honk!" The car horn blared in the distance as I searched for "Indian beads" on the sand bar. Hickory Creek was flowing nicely…a steady, shallow stream of glistening water wandered back and forth across the wide, sandy creek bottom. I wasn't ready to leave…I wasn't ever ready to leave this place.

I knew the next time I came to the creek the sand bars would be rearranged by the wandering stream. Clumps of leaves may be hanging over my head on twigs that caught them as floodwater washed them by. Sticks, maybe even logs, would appear where none had been before…washed downstream by a torrent of muddy, swirling water washing off the nearby farm fields. New, interesting rocks would be exposed…maybe some mussel shells, an old bottle, pieces of a crock, or "Indian beads".

I had just found an old unbroken glass pop bottle and thought it would be handy to put "Indian beads" or interesting rocks in to take along. Glancing around quickly for Indian beads, I saw none…but a few small rocks caught my eye. I put a few in the bottle, then tried to put one in that was a little too big. Trying to cram it in, I ended up wedging it tightly in the opening of the bottle. I couldn't push it in or get it out. "Honk! Honk! Honk!" Three honks! I knew I had to get going now. Frustrated, I dropped the bottle back on the sandbar and ran back to the car. Little did I know, I would find this bottle again forty years later…at the perfect time.

When I was a young boy, going to the "crick", as we country people say it, was my favorite thing to do too. Grandma would take my cousins and me a few miles south of Brownstown, Illinois to Little Hickory Creek, Little Sandy Creek, and sometimes Flat Creek. Up there, the creeks are sandy and it takes little imagination to feel as though you are at a beach.

When we were small, Grandma would climb down the bank and walk the creek for a while with us, then sit with her feet in the water. I never got tired of playing in the giant "sand pile" that the sandbars created; building little dams and ponds for the clear, flowing water to fill up and wash away. As we got a little older, she'd stay in the car and take a nap while we played. Catching crawdads, tadpoles, frogs and minnows was great fun. I learned to skip stones, and while looking for the best skipping stones, we would find fossils of seashells and "Indian beads" (which were really crinoids, ancient marine worms). As we got older, she'd just drop us off at the bridge and say "I'll pick you up at the next bridge that way" and point, adding "When I honk the horn, you come." This way, she could go run some errands without kids tagging along.

Looking back, I guess it's a wonder nothing ever happened to us while we played for hours in the creek. But, this was back when kids would roam around town by themselves all day on their bikes. Parents would just say "Go outside and play" and "Be back by dark!" and that was it. We were free to explore and use our imagination.

According to Richard Louv, author of "Last Child in the Woods", children today are suffering from what he has termed "nature deficit disorder". According to Mr. Louv, "Nature deficit disorder describes the human costs of alienation from nature, among them: diminished use of the senses, attention difficulties, and higher rates of physical and emotional illnesses."

For too many children today, Louv's assessment is far too accurate. Despite the popularity of all things "green", and supposed heightened awareness of the environment, the environment is to many young folks today just something they see on television or a computer; not real life. Sadly, "Our children", he writes, "are the first generation to be raised without meaningful contact with the natural world."

Walking trails is fun, and is a great way to get out and see nature. But, sometimes it's good for kids to explore water, mud, rocks, etc. They love places where they can find little critters like tadpoles and crawdads. This type of excursion promotes unstructured "hands-on learning" that is so important in childhood development.

I took my kids to play in the creek too. Here, the creeks are rocky, so we wore water shoes to protect our feet. I stayed with them and didn't just drop them off. Walking a creek can be a lot of fun for the kids and adults alike. Dogs love it too. So, let's go to "the crick"! And, when we get there, remember to let the children experience nature… to get in the water, get their hands dirty, handle the harmless little critters they find, build with rocks and sand, skip stones, etc..

Yes, clothes may get dirty, knees may get bruised, we might get mosquito bites… but that's part of the experience. Try not to ruin it with too much protectiveness and caution. Seriously, parents, did we ever use hand sanitizer when we were kids? Insect repellant? That stuff was for sissies when I was a kid. Sunscreen? Hadn't been invented yet. Sure, take precautions, provide supervision, but remember to let the kids be kids. I think our boys' favorite thing to do is wade in a creek or pond looking for salamanders, frogs, crawdads and minnows. These outdoor excursions were the first of their "journeys".

My journey to becoming the tenth Berea College Forester was not just a physical journey of moving geographically, but a life journey. We need not move locations to embark on such a journey…but it helps.

An odyssey is an extended, wandering journey. The word comes from an ancient Greek poem, by Homer, about the wanderings of the hero Odysseus during the ten years after the fall of Troy. To make a long story short, at the end of a tumultuous, dangerous journey back to his homeland, Odysseus is finally reunited with his wife and son.

A key portion of the story consists of Odysseus' wife, Penelope, testing him to see if he is really who he claims to be, by telling him that her bed has been moved. Only she and her husband could know that this was not possible, as their bed was made from a living and still rooted olive tree. In the book, we read that the olive tree was "growing strongly in the courtyard, and it was thick, like a column" The column-like trunk of the olive tree formed the center structure of the house as well as one of the bedposts.

The olive tree, a major source of food, oil and wood for the Greeks, was held in great reverence. The use of the tree in this way symbolized the strength and protection of the husband and the fertility of the woman. It also paid homage to the tree and the land that supported it and the family. It is the picture of "husbandry"; which used to be synonymous with farming.

A little research into ancient Greek and Roman agriculture reveals that ancient Man knew a great deal about cultivation of olive trees and what we would today refer to as "sustainable living". For centuries, the arid, desolate eastern Mediterranean region was assumed to have a much drier climate than in ancient times. This was thought to be the case because much of it is a virtual wasteland today; while in ancient times it sustained vast olive, almond, pomegranate, fig and date orchards as well as vineyards and grain fields.

Recent research, based on the study of ancient surviving trees, has revealed otherwise. The trees testify that there was no more rainfall in ancient times than today; and that the cyclonic spring storms; followed by drought in the summer, has been constant for thousands of years.

The difference is that the ancient agrarians left half of the land fallow every year to conserve moisture and rest the soil. They also collected rainfall for irrigation through an elaborate system of terraces, depressions, underground tunnels, cisterns, and aqueducts. Some of the tunnels were bored through entire mountains entirely by hand. These tunnels still baffle engineers today as to how they could bore from both sides simultaneously and meet precisely in the middle.

Unlike modern irrigation, which wastes most of the water through evaporation, ancient practices kept the water underground where it did not evaporate or cause soil salination.

These ancient agricultural methods of water and soil conservation were abandoned after the collapse of the classical civilizations. Consequently, the land has degenerated to the present condition and desert continues to expand in arid regions throughout the world.

Hopefully, Man's odyssey… our extended, wandering journey… will bring us back to sustainable practices and a reverence for the land epitomized by Odysseus' olive tree.

The symbolism of the olive tree with sustainable agriculture, marriage, and home is not an accident.

For many young people, going to college is a major journey…and it can change the trajectory of their life forever. I do not believe college is for everyone…especially if it means borrowing large sums of money that could take many years to pay back. For many, simply getting a job or going to a trade school, the military, or an apprenticeship program could be a better fit. Starting a business is a great fit for some too.

Many times over the years, I have even wished I hadn't went to college myself. Why? Don't I have the "dream job" I always wanted? Yes, I do. But, what I REALLY wanted was to have my own forest to live on and manage as my livelihood; preferably near where I grew up so I could maintain close relationships with my extended family. If I had pursued my ideas and passions in an entrepreneurial fashion instead…and stayed in one place long enough to achieve them…instead of bouncing around in pursuit of the perfect job… I am confident I could have accomplished that by now, as every move resulted in financial loss and loss of assets that simply staying in one place would have prevented.

Nevertheless, for those who do choose to go to college, the problem of student loan debt is not an issue at Berea College, as it has been tuition-free since its founding. Berea College is a "work college"; one of just a handful in the nation. Students have to work ten hours a week while attending College… but they gain invaluable experience through their work…and this seems to help many of them in the workplace after graduation.

The best thing about Berea College is…to my way of thinking…is not that it is tuition-free, but that it has "The Great Commitments". The Great Commitments were originally adopted by the General Faculty and the Board of Trustees in 1969; they were revised and similarly approved in 1993 and most recently in 2017. I present them to you below:

1) Educational Opportunity: To provide an educational opportunity for

*students of all races, primarily from Appalachia, who have great promise
and limited economic resources.*

2) *The Liberal Arts: To offer high-quality liberal arts education that engages
students as they pursue their personal, academic, and professional goals.*

3) *Christian Values: To stimulate understanding of the Christian faith and its
many expressions and to emphasize the Christian ethic and the motive of
service to others.*

4) *The Dignity of Labor: To promote learning and serving in community
through the student Labor Program, honoring the dignity and utility of all
work, mental and manual, and taking pride in work well done.*

5) *The Kinship of All People: To assert the kinship of all people and to provide]
interracial education with a particular emphasis on understanding and
equality among blacks and whites as a foundation for building community
among all peoples of the earth.*

6) *Gender Equality: To create a democratic community dedicated to education
and gender equality.*

7) *Supportive and Sustainable Living: To maintain a residential campus and
to encourage in all community members a way of life characterized by
mindful and sustainable living, health and wellness, zest for learning, high
personal standards, and a concern for the welfare of others.*

8) *Serving Appalachia: To engage Appalachian communities, families, and
students in partnership for mutual learning, growth, and service.*

Achieving these commitments has been, and continues to be, a journey for
the institution and for the individuals who work and are attending college here.
There is a term that is sometimes used at Berea College called "Being and
Becoming" that speaks to this process.

My position as College Forester places me in the category of "staff". I am
not a faculty member. So, my role in the first two Commitments (Educational
Opportunity and the Liberal Arts) is limited to a supportive nature. Managing
the forest in a way that enhances educational opportunity is my main function
in these areas. Since I am a Christian myself, I fully embrace the Christian

Values Commitment; but those who are not Christians are welcomed by the College too…and can focus on the motive of "service to others" with whatever faith, or no faith, they themselves identify with.

The Dignity of Labor, The Kinship of All People, and Gender Equality all go hand in hand, in my experience. I have found that folks working together… whatever their background, religion, race or gender…find kinship more readily than in any other endeavor. I will cover this more in the next chapter…the Work chapter.

In my position as College Forester, the Sustainable Living Commitment and the Serving Appalachia Commitment are my focus. I can try to manage the forest in a way that is conducive to these commitments, but how do I "… encourage in all community members a way of life characterized by mindful and sustainable living, health and wellness, zest for learning, high personal standards, and a concern for the welfare of others"? And, in Appalachia, where Berea College is particularly engaged, how do I "…engage Appalachian communities, families, and students in partnership for mutual learning, growth, and service"? This takes interaction with people and with the community. Communication is key, and I hope this book helps in that regard as ultimately it would be nice to have an influence not just in Appalachia, but beyond.

———————

The clear slab of white oak was thirty-three feet long. It still smelled fresh and felt damp…its 110 years of growth rings exposed to the sun. Beautiful, blonde, water-proof, and absolutely clear of knots, it was amazing to see. I hoped it passed muster. I hoped we could saw enough of them to attract the shipwright to fly all the way from Boston to Berea to inspect them. Hopefully, the plank would spend the rest of its existence floating on the sea as part of a historic wooden ship. Ernie had ended up driving all the way back to northern Ohio and back again to get the slabber part for his Lucas sawmill. Holger had to go find another key for his borrowed bulldozer and try to roll and drag the massive logs into position. Glen and I had been out there wrestling with the sawmill's extended framework; chopping off muddy bark; and moving slabs around for nearly twelve hours straight as we tried to achieve what no one else had apparently attempted to do: supply knot-free white oak planks for the repair of the Mayflower II ship.

Those brave people who arrived on the original Mayflower took a journey to America that changed their lives forever, and changed the course of this nation as well. I was proud to be involved with a project at Berea College that connects us to this historic journey in a contemporary and exciting way. In 2016, Berea College harvested nine white oak trees and sawed them into thirty foot long planks to contribute to the restoration of the Mayflower II ship. The original Mayflower sailed to Plymouth in 1620 and no longer exists. However, a replica was built in England in 1957 and sailed across the Atlantic. It is operated by Plimoth Plantation, in Massachusetts, and is a popular tourist attraction. While not nearly as old as the original ship, it is beginning to need a major overhaul. Obtaining long, clear plank wood and curved futtock wood has been a challenge for the shipbuilders. They had been searching for two years, without any luck, to find white oak trees suitable for the restoration.

I was able to find several white oak trees that had the characteristics necessary to meet Plimoth Plantations specifications and decided to try to become the supplier for the Mayflower II restoration. The trees had to be large enough and tall enough to produce thirty foot long clear planks four inches thick and twelve inches wide; not counting the sap wood. While many large trees can be found with a thirty foot straight log in them, it almost unheard of to find clear wood all the way to thirty feet. Yet, nine trees were located and successfully sawed into the material they needed. This provided enough material for the shipbuilders to get started, but not enough for the whole project.

The sawing of the thirty foot logs took specialized equipment and was extremely labor intensive and time consuming. I told the Plimoth Plantation folks that we just didn't have the time to saw logs for them again, but we would be happy to put together a truck load of logs to sell them if I could find enough to make it worth their while to come and get them. They have their own means for sawing logs out there, at the shipbuilding facility. So, as I marked the next harvest for Berea College, I kept an eye out for suitable white oak for them. I was able to locate three more of the very tall, straight, clear logs needed to make knot-free thirty foot long planks. Even as big as these logs were, that was only half a load.

The ship's captain, Whit Perry, had told me about how the planks have to be bent to attach them to ship, and how they are secured to the futtocks, or ribs of the ship. The futtocks are thick pieces of curved wood. Ideally, they are cut with the grain from logs with natural curvature. This makes them much stronger than if the log was straight and the futtock had to be cut across the grain to obtain the correct shape.   I had this in mind was I searched for tall, straight white oak. But, it seemed to me that we had been "lucky enough" to find such ideal straight, tall white oak… surely, there weren't going to be huge white oak with natural curves in them too. Luck must have been on our side, because I found three highly curved, sound, huge, clear white oaks in the same planned harvest area. Just the right amount for a full load… a VERY full load.

Monday morning, the loggers were able … after several attempts… to successfully load all six of these huge logs onto a truck bound for Mystic Seaport, Connecticut, where the Mayflower II will undergo its restoration after the end of the tourist season. This was where my mom and I had went when I was fourteen and stayed with the elderly lady who lived right on the coast. She had shown me pictures of her beloved ancient white oak that had once stood there and I couldn't help but connect this memory with the opportunity to supply white oak planks to the nearby shipyard for the Mayflower II reconstruction. The wood would "live on" as a historic ship that would tell not only of the history of the Mayflower, but of the size and quality trees must be allowed to attain to build such a wooden ship. Luckily, the guys got the logs loaded before another one of those downpours hit and the trucker missed a tornado as he passed by Grayson, Kentucky headed east. One more worry… that the truck would be way overweight at the scales, turned out OK as well. Another trip accomplished in this chapter of journeys.

Some might ask, and indeed some have, whether that reconstruction project really justified the sacrifice of those mighty old friends in the Berea College Forest. For me, and this is how I answer, mankind has been using forest products for innumerable purposes for time immemorial. This use seemed to have the potential for something more, a reminder to all who encounter the story, or eventually see the restored ship itself, of all those uses, and how recklessly humans have pursued their ends at the cost of entire forests.

Probably everyone has heard that tall, straight, clear white oak is valuable. Generally, it is used for veneer and sold as such. However, it's pretty unusual to be able to sell a crooked log for a high end use that makes it worth trucking all the way to Connecticut. Yet, this is what was done. The revenue received from timber sales on the Berea College Forest has always gone to help support the College, and in my time here has gone back into further investments in the forest. I definitely hope to get a chance to go out to Massachusetts there and see how they used our wood in the reconstruction and eventually see the ship sail.

Our involvement with the Mayflower II project would result in us being able to also work with the Western Flyer ship's reparations five years later.

The word "sustainable" has become somewhat of a "buzz word". It's great that "being green" has become popular. Or is it?

Country music singer Barbara Mandrell had a big hit in 1981 with her song

I was Country When Country Wasn't Cool. I remember hearing that song and thinking "Me too!" Not too many kids my age liked country music, but I was always getting "dragged along" to square dances and fiddle contests with my grandma. I really liked it. The only negative thing was that the old ladies liked to squeeze my cheek or rub the top of my head when Grandma introduced me to them. The first one of these "gigs" I was dragged to, as we walked in the door an old man was playing a fiddle to sound like a chicken clucking! I was hooked from day one!

Grandma and Grandpa often had "Country & Western" "shindigs" in their basement. "Old timers" would come from all over; bringing their fiddles, banjos, and guitars. They would stay up late into the night, and many a time I'd fall asleep to their lively tunes and foot-tapping. Sometimes, my cousins and I would dangle a fake spider from a string down through a hole in the floor to get a reaction. Someone would play along and pretend to be scared.

By the mid-80's, Randy Travis had firmly established country music as cool...even among teenagers like me. Later, Garth Brooks really made it cool. However, as more and more people my age began to jump on the country bandwagon, I didn't have much use for it anymore. "It's just a fad now... not the real thing", I thought to myself. I rarely listen to what passes for country music these days...although I do make a few exceptions. Nowadays, "Going Green" is becoming "cool". And, that's "music to my ears". However, many of us were conserving, planting trees, and considering our environmental impact long before "eco-friendly" or "carbon footprint" were part of the vocabulary. The real thing, when it comes to "being green", seems to follow the pattern country music has...and become more of a fad than authentic.

My cousins and I would roam around rounding up glass pop bottles to return to Gene's Red Fox for money. "Why would anyone throw them out?" we wondered, as we counted out our change to buy candy at the store. Then, we'd go next door to the hardware store and buy some fishing hooks and bobbers. Now, both stores are out of business and you have to drive to the next town (past discarded plastic bottles) to go to a store. Grandpa always wrapped up his sandwich to take to work for lunch in a barely big enough piece of aluminum foil. He was so averse to wasting anything that he would carefully unfold the aluminum foil and then fold it back up again; putting in back in his lunch box. He'd use the same piece of foil for weeks, until it fell apart.

We hear about how we should upgrade our appliances and our heating and cooling systems to be more efficient to conserve energy. Good idea. But, few people then had air conditioning, and the clothes dryer was solar and wind powered... a clothes line. Zero energy use.

"Conserve water… turn off the faucet when brush your teeth, and put in a low-flow shower head", they say.   As I recall, Grandma's house had a cistern to collect rain water for watering the plants and washing the car. And, as far as bathing went, several of us would take turns bathing in the same water. That sounds gross now, but was common practice then. And, if we had dared to leave the water running while brushing our teeth, someone would holler: "Quit wasting water!"

I'm excited, but at the same time wary, that "going green" is becoming the cool thing to do. You see it on bumper stickers, tote bags, and on the television every day. Really being green, I think, is a lot more than those easy proclamations. It has more to do with personal responsibility than anything else. It's not just a fad; it's a way of life. Often, it means not buying something at all, rather than buying something advertised as green. It means, too, holding elected officials accountable for the society-wide and even world-wide choices that need to be made for the sustainability of our planet.

So, were you green when green wasn't cool? If the coolness of being green wears off, I hope we continue to live the green lifestyle anyway… it just makes sense and it may be the most important thing we can do for the future world in which our children and grandchildren will live.

"It may be that when we no longer know which way to go that we have come to our real journey. The mind that is not baffled is not employed. The impeded stream is the one that sings."
– Wendell Berry

I find it curious the "zest for learning" is in the Sustainability Commitment. I seems like it would go without saying that a "zest for learning" would go along with either the Educational Opportunity or Liberal Arts Commitment. I was on the Strategic Planning Council at Berea College when we discussed rewording the Great Commitments, and was glad that this zest for learning part was kept in the Sustainability Commitment.

Last week, I was driving a team of Suffolk Punch draft horses; pulling out an ash log that Ben Burgess had cut down the day before. His dad, Gary, was riding on the "arch"…a two-wheeled cart used for holding the forward end of the log off the ground…with me. I was getting some hands-on training from Ben, but his dad is one of those guys who can do anything…he started a tool and dye business that was very successful, had drilled oil and water wells, and has forgot more than most people know. He smiled at me as I was concentrating on what I was doing and said "Learning is something we never get done doing, isn't it?"

Gary had went over to check on his tapped maple trees and thought I could use some coffee. He handed me his old Aladdin thermos (like mine, made in Nashville instead of China like they are now)…saying "A rabbit hunter lost that over in my woods…found it a while back…pretty good thermos, huh?" His wife had brewed the coffee…hazelnut…and the smell of that coffee, the freshly cut logs, the horses, and the hay field all blended together to create a wonderful aroma as we bumped along on the arch cart.

"Get an education." Well, how do we do that? We can go to college…and perhaps we should…at least some of us. But, like Gary said…learning is something we never get done doing. That day, I learned things that no textbook could have taught me. I learned how to read subtle signals the horses made between one another; and how to react appropriately to maintain a calm control of the situation. I learned how to beat the "grabs" into the end of a log so they wouldn't pull out…and how to knock them back off when the log was where we wanted it. The grabs and the tapered sledge hammer designed to knock them out were specially made since such horse-logging items are not in common use anymore and can otherwise only be found at antique stores and auctions. The logging arch cart was built by Ben and his dad themselves.

The day before, since it was raining all morning, we had intended to trailer the horses to an Amish farrier to have their shoes refitted. But, he had hurt his back and was out of commission at the present time. I had been there before, and was disappointed we couldn't go…as well as concerned for him having a hurt back. A horse-shoer needs his back…but certainly uses his brain just as much. With only an eighth grade education (formally), he impressed me…as do a great number of Amish that I have met…as being not only one of the most knowledgeable people about horses and mules I had ever met…but also a keen businessman, entrepreneur, and mentor.

Some of the smartest, most resourceful, and "greenest" people I have met are "salt of the Earth" folks like Ben, Gary, Amish businessmen and women, and horse and mule loggers I have met while learning to do "horse logging". Jason Rutledge, the Virginian who founded Healing Harvest, would not like me to even call it "horse logging". He does not claim to be one…even though he is one of the best. His is, rather, a "biological woodsman".

What is a "biological woodsman"? Well, I could ask Jason what his definition is, but I think just observing how he and his apprentices operate provides the best idea of what it is. Gary owns over 1,000 acres of forestland. He and

Ben work the land themselves with horses…cutting the worst trees first. Ben makes a living doing this. I am glad they didn't get an MBA at some college, or they might have been talked out of it. I can't imagine any financial adviser telling them to do this…it's so slow they will never get done harvesting their timber! They could surely make more money selling it all at once to be logged by conventional means…invest the money…and wouldn't have to do any work!

My Grandpa Torbeck, who turned me on to reading Goethe, would always say "When you make a decision, always ask yourself: 'If everyone did this, would it make the world a better place or not? If the answer is yes, do it. If the answer is no, don't do it." So, if everyone threw their wrappers out the window, all the roads would be lined with trash. If everyone sold off all their timber at once, and invested the money, there would be no timber for a long time. You get the idea. A "biological woodsman" makes decisions based on what is going to make their woods a better woods in the long run; not just the immediate "bottom line".

"Do you know what species of trees those are?" the forestry professor asked me as I looked up at a grove of towering evergreen giants. Every tree was evenly spaced, of exquisite quality, and looked the same size and age. Fortunately, I had done a little bit of homework…so I knew it was a Douglas fir (*Pseudotsuga menziesii var. menziesii*). He took out a notebook and as read from it: "These trees were planted in [about 1860] and it was pruned in [several times since planted] and thinned in [several times since planted]."

"Wow! You have records from that far back?!" I exclaimed as I wondered how the trees were pruned to such a height. "Fortunately, they were not lost during the war, and these stands survived as well", he explained, then went on to describe how the trees in the Rottenburg Forest were kept track of like this so that when they are offered up for sale, the timber buyers are presented with a guarantee of clear wood to a certain depth and volume in the tree…which can be accurately assessed with these records.

"Well, so how do they prune them to such a height?" I asked. "Oh, our foresters carry an extension ladder from tree to tree and prune them by hand… when it is necessary to prune higher than the ladder, they raise the ladder another length and tie it to the tree…we go to about ten meters." Everyone marveled that this was done…and wondered why it wasn't done mechanically. "Oh, it was determined that only pruning by hand could achieve the quality of job." Oh, ya…of course. Hmmm…

He went on to tell us how Douglas fir became a major timber tree in Germany. A German forester went to Oregon before the virgin Douglas fir forests were cut down and located the best trees; obtaining seed samples from them and bringing them back to Germany. The trees were discovered to grow extremely well there; growing faster than their native firs and pines. As the forests were cut down in the Pacific Northwest, where they are native, timber companies started to look for superior seed sources and found out that the German forester had been way ahead of them…seeds obtained from superior trees had already been planted in Germany and were grown into seed-bearing trees already. "Oh yes, we actually sell a lot of seed to the USA for planting. In fact, the tallest tree is Germany is a Douglas fir." Hmmm…

Everywhere we went on the community forest, there were walking trails and beautiful wooden interpretive signs explaining whatever forest management was going on at different locations. Frequently, people walked by and greeted us. I could read enough German to make out that some of the signs forbid entry by vehicle, but there were no gates whatsoever on the forest. "How do you keep people from driving on the forest roads where vehicle access is forbidden?" I asked the forester. He pointed to a sign "Oh, it says right there that vehicles are forbidden, so no one drives there." How interesting, I thought, as I recalled all the instances where this does not work on our forest in Kentucky. Even gates commonly don't work…getting tore down or even stolen.

We came to a place where a recent harvest had taken place and some of the logs had still not been hauled away. The logs were mostly Sessile oak (*Quercus petraea*) and had been under management for over 200 years…and were awaiting transport to France where they would be made into wine barrels. The harvest was obviously very recent, but there was no evidence of logging to be seen. No ruts, not skinned up trees. So I asked "How did they get the trees out?"

The forester explained that equipment is not allowed on the forest, as it would compact the soil…so the equipment must stay on designated forest roads and the logs have to be winched out with long cables. A person tends to the cable so that it does not skin up any of the trees. The forest roads seemed far too far apart for this to be practical, or even possible, but it was how they did it. A lot of small trees were marked…too small to be of merchantable value…and some other foresters were on the site cutting them down.

The foresters knew we were coming, and stopped to take a break and talk to us. Their truck was nearby, and like my work truck, it has a tool box across the back. They went to one side of the toolbox, opened it, and got out a saw file to sharpen their chains with. Then, they went over to the other side and took out a wooden box…sat it on the tail gate…opened it up…and asked us we would like to have a beer or home-made schnapps?!

Well…I learned that this was perfectly normal to drink on the job there. "Oh yes, don't you allow it in America, 'Land of the Free'", one of them said, with a wink and the other told us how there are beer machines, like pop machines, at the factories too, and even at the school cafeteria beer is served. The forestry professor laughed and told us how parties were encouraged at the college to help form close comradery among the students, and free beer was provided on the weekends to encourage it. Usually, they also roasted a wild boar…the forestry students were encouraged to hunt, too…to help keep the wild boar and deer numbers under control. Wild boar was particularly damaging to the forest, and hunting season was year-around, with no limit…and the college even had a cleaning station for the students to use to dress the animals…plus, they could sell the meat to local butchers and it could be found in any of the local restaurants. Students were required to build their own hunting stand to hunt from… somewhat akin to our deer hunting stands here, but permanent and built from thinned trees from the forest. These stands were sometimes located in rather urban settings, and I noticed one a short distance from a *Bierstube* (beer garden) amid a residential neighborhood.

"Wow, I saw one of those stands right on the edge of Freiburg across the road from a *Bierstube*. They allow them so close to homes and businesses?" I asked. "Oh, yes, that's probably where the deer or wild boar were at, so why not? We use silencers on the guns so the shooting does not disturb anyone. It would be rude to make a loud noise, you know." "What?! Silencers are allowed?!" I marveled, and asked about safety. He went on to explain that to become a Jaeger (hunter) you had to complete a wildlife management course and then every year pass a "hunter's cinema course". The course consisted of real-fire shooting of targets on a movie screen…with wild boars and deer coming out along with people in various scenarios. Hunters must shoot about 80% of the animals, with good clean "kill shots". If they accidentally shoot a person during the cinema course, they are disqualified that year from hunting. Their state, Bavaria, also had a phone app that the hunters were required to use so they could log into which stand they were hunting, and all other hunters could see on their phones where hunters were located and thus know if the stand was occupied and thus take extra and safety precautions. Wow.

"Don't you have any problems with students having access to alcohol?" I asked…thinking how so often when I come across a deer stand in the USA there are beer cans littering the ground around it. The forestry professor gave me a funny look like "Why would there be any problems?" and said "Of course no one would drink while hunting. But, oh, I can remember one time. I think it was about nine years ago. Two students were late for class the next morning after one of our college beer and wild boar parties. We allowed them to finish the course, and then when they received a failing grade, we reminded them how they were late that day so their class was incomplete. We never had any more problems." "That's it? Seriously? No drunk driving?" "Oh no, of course not! Who would be that stupid to drive drunk?! I never heard of it."

"So, let us tell you about what we are doing here", said one of the foresters as we finished tasting their home-made schnapps. He explained how the small trees were being designated to thin the stand, to make room for the better ones, and how these would not be wasted, but used for firewood. "By the weekend, people will have them cut up and taken away for firewood", he said. "But they are not allowed to use equipment, right? How do they remove the wood?" I asked. "One group has horses they use, and you will seem that too…but mostly families come together and they carry it out by hand. There might be thirty people in a line, handing it one to the other, to remove it…it is great fun… families do it together", he said, with a chuckle.

Wow. Just wow. Things really were done differently here, to say the least. My trip was definitely living up to what I had hoped to see…here was a whole society acting like the "old timers" I had been fortunate to work a few years with in rural Illinois.   So, how did this trip come about?

In early 2016, I got to plan a forestry-focused fact-finding trip to Germany for a group from Berea College to take. This was like a dream come true for me, and it came about amid a discussion with Berea College's President, Lyle Roe- lofs. I had inherited a forest management plan from the previous forester and administration, but I had my own ideas I wanted to implement going forward, and felt that this was the direction that was more appropriate for Berea College given our mission and our strategic positon as an educational institution. Namely, I wanted to seek a more "nature-based" forest management that would incorporate using horse logging in some fashion, and also to try to foster community-forest elements that would hopefully promote a better "land ethic" here.   Lyle asked me where I would suggest learning how to implement these ideas, and I said Germany. To my astonishment, he told me to prepare the itinerary. I didn't know he had studied in Germany years ago and had a lot of the same ideas.

The trip took place in September 11-22, 2016. Our first destination was Freiburg im Breisgau, on the edge of the famed Black Forest.   Freiburg is often touted as one of the "greenest" cities in the world, and it did not disappoint. The beautiful old city center, the Münsterplatz, surrounds the immense Freiburg

Münster…the great Gothic sandstone cathedral built around 1200. There was a huge farmer's market set up around the Münsterplatz, with home-made breads, cheeses, sausage of every kind of meat from duck to wild boar, flowers, wooden carved cuckoo-clocks, food and beer stands. Everything organic, everything locally grown or made. I soon realized that finding organic, locally grown food was the norm here, not the exception, as it is in America.

We toured the Freiburg Community Forest…one of the largest and oldest in Europe. Our tour guide, Andreas Schäfer, was exceptionally knowledgeable and interesting. He told us first-hand accounts of the tense situation he faced as a West German soldier stationed in West Berlin during the days leading up to tearing down of the Berlin Wall. He related how he decided he would not fight the East Germans…his fellow countrymen separated by the Iron Curtain since WWII…if war broke out. He told of how his grandfather never returned from WWII, and how after the war, the French forced them to cut down nearly all of the forests in the region and how everything we can see today had to be replanted and rebuilt.

Grimm's Fairy Tales came alive to us in the forests, as we saw wood carvings of huge mushrooms, various fairytale characters, and little huts throughout the forests along trails. We toured a huge, beautiful community garden area where people maintained gardens for both food production, and flowers everywhere beyond the imagination. We rode up to the top of Schauinsland Mountain on the cable car and climbed the tall tower at the top to look across the beautiful landscape. There were wind mills in view, but not obstructive. Many, many homes and barns had solar panels on their rooves. I wondered how a place could come to be ran so well…to achieve this status of such a "green" city. They even had a mayor who was a member of the Green Party, and there was even a "Mayor of the Environment". Imagine that!

We met with the Mayor of the Environment's chief of staff, Franziska Breyer. She told us how her city had gotten to where it is. She explained how in the 1970's, a nuclear power plant was designated to be built near Freiburg. The citizens did not want it…and they had to work together to stop it. They realized that just to keep the nuclear plant from being built "in their back yard" was not a solution…they had to come up with an alternative, so that it didn't need to be built…anywhere. Alternative energy development, along with conservation efforts, had to be implemented. Businesses, churches, environmental groups, and citizens of all walks of life joined together to make this happen…and it resulted in one of the greenest cities in the world…which in turn has made Freiburg have a booming economy as well. People want to travel here and they want to live here.

Next, we traveled to the World Heritage/UNESCO Biosphere in the Regional Natural Park of Ballons des Vosges in Alsace, Lorraine, and Franche-Comté. Here, we saw some beautiful, biologically diverse, "wild" forests that contrasted with the highly managed forests in Germany. This area was incredibly beautiful, and we got out and hiked on some logging roads and toured their visitor's center.

Next, those of us directly engaged with forest and land management... myself; Glen Dandeneau, the Assistant Forester; Bob Warren, our Forest Technician; Sean Clark, Agriculture professor at Berea College; and Jason Rutledge, a renowned "biological woodsman", Suffolk horseman, and head of Healing Harvest Foundation...went for five days to Rottenburg am Neckar. Jason went along to help us to evaluate animal-power applications. Here, we toured the Hochschule für Forstwirtschaft (technical college of forestry) and the community forest the college uses for its training and research. This is where we really got to experience how forestry is done there. Also, since I and Bob stayed with our host, Stefan Ruge (Botanik, Waldbau-Grundlagen; Botany and Forestry professor) at his home, we had such an opportunity for interaction. We even went wild boar hunting with him and his son one evening. I sat with his son, and although we saw no wild boar, he optimistically told me about how he was on track to become the eighth generation forester in his family. That's right... eighth generation. We stayed with Stefan and his wife, met their two grown sons and daughter, and had the best time.

After the trip, I realized a few more things that were different there. The Ruge's treated us one day to a boat ride on the Neckar River at the ancient, walled city of Tübingen. Everywhere, students were sitting on high stone walls along the river, reading or working at their laptops. We got on a boat, and while there was a bit of instruction which amounted to something like "Don't stand up", there were no signs like in America warning people to do this or telling them to not do that. I also realized that, while I had driven the rental van for the entire trip, I had not seen a single police car...and only one policeman...on a bicycle! Maybe this was because there was no speed limit on the Autobahn. But, if you run a red light, I was told, the intersections have cameras, and you will get a ticket in the mail...three times and you lose your license...I think, for a year. It took some getting used to being back in the USA...it felt like I was being harassed all the time until I got used to our different way of doing thing again. "Oh ya, we are treated like toddlers here...consumers and potential criminals...not responsible, thinking adults."

Then, I thought about the students there...college was free. There was universal health care that was almost free. Literally no one was obese, or even overweight. The drug store I went to get medicine to help me overcome jet lag, was tiny and had almost all herbal, over-the-counter products. The areas with the prescription drugs, behind the counter, was the size of a walk-in closet. This contrast really hit me when I went through the drive-up window of one of the six pharmacies in the small town I live in, to pick up my eye drops (for dry eyes…they haven't been producing enough oil content in my tears). I pulled through and parked to use my phone before driving off. A police officer was at my window within five minutes. Seeing that I was not getting high off whatever I picked up at the pharmacy, he explained that he regularly patrols the pharmacies since there are often people passed out in their vehicles from this…and this was the same row of businesses where all the air conditioning units were stolen the year before as well…trying to make sure this didn't happen again.

"How did we get to this point?" I thought to myself…looking around at all the trash in the parking lot and thinking it's not just a "land ethic" we are lacking, but a "humanity ethic". Clearly, changes need to be done to improve our care for the land, for our forests and environment…but for this to happen, people need to be healthy, educated, fit, and have a sense of hope and purpose. Traveling abroad gave me more than a sense of direction for managing forests; it gave me a new sense of why so many people here are so often chasing dollars or the next high instead of engaged productively and passionately in some worthwhile effort with each other.

———————

King Bolete *(Boletus spp.)*…white mushrooms…sure seemed to be "king" in Ukraine in late September. Nearly everyone who got on the train as we traveled through the forested areas, boarded with a large basket or two of these delicious mushrooms. They look like the perfect mushroom you see on the cover of an Allman Brothers album…and Ukrainians widely use them in cooking…along with their delicious local cheeses. At each stop the train made, there were people setting up to sell the mushrooms…along with all manner of other local produce, crafts, woolen clothes and blankets, and other locally produced goods.

Traveling across Ukraine by train…old, bumpy, wooden-seat Soviet-era trains…I wished we still had this manner of transportation widely available in America. Berea, like every town I have lived in, has an old train depot that trains have not stopped at for decades. But, here in Ukraine, the train stopped at every little "whistle-stop" to pick up and drop off passengers. It was interesting to watch them get on and off. I would have been surprised to see young children traveling alone, it I hadn't already seen them walking alone all around busy Kiev days before. There seemed to be much less sense of danger here in children getting around alone…even in the large, capital city…than I was used to seeing even in small towns in America. I was surprised to see so many dogs walking with people too…almost always off leash, but perfectly behaved, healing beside their owner as if on a leash.

Everyone…was quite fit, dressed very nicely to be out and about. I had heard how there is such poverty here, and it was obvious that few people owned cars…elementary schools we toured still had outhouses (outdoor toilets) and burned wood in old pot-bellied woodstoves…the children taking turns doing chores. The buildings were old and weathered and the classrooms still had the old chalk boards…an old wooden piano…hard wooden chairs…also stylized Orthodox Jesus and Mary pictures hanging on the walls. The children behaved like they used to in America too…lining up in single file and quietly paying attention; raising hands to answer a question; replying in unison when their teacher asked them to repeat something. Rushing outside for recess with excitement and smiles, they seemed to have it better…in my way of thinking… at least at this stage of their lives…than I felt like many American children do. But, the teacher said that many of their students, as they get older, get hooked on alcohol…at an increasingly young age…and more recently, also drugs.

Looking out the train window, I was astounded to see cattle walking along beside people, without a lead rope, just as the dogs had done in Kiev. One old man was riding an old bicycle, and a big milk cow plodded along beside him at a good pace on the road shoulder as fast-moving traffic whizzed by. I was told that many people tended to cattle, sheep and goats in the unfenced common meadows…taking them out in the morning and back at night. The Soviets had kicked the kulaks and farmers off the land and property ownership in most places was still in a state of limbo…with people owning their homes, but the surrounding pasture lands ownership not settled.

Remarkably, the Hutsul people…traditional high altitude mountain people who grazed their livestock in the high meadows in summer and came down to live in their villages in winter…had largely survived as a people through so many tortured centuries of invasions, wars, and communist take-over. An old Irish- man I met in Ivano-Frankivsk…who had hitchhiked to the Carpathian Moun- tains from Ireland and loved it so much he stayed…said he knew this was "home" when he met the Hutsuls. He was a stranger and didn't even know the language, and yet as he wandered into a mountain village, he was offered an unused cottage to live in for the summer months when the inhabitants did not need it…teaching English and playing his violin for food, he said he had travelled the world and never met such giving and trusting people as he had found in Ukraine. He lived there long enough to learn Ukrainian well and then moved on to Ivano-Frankivsk…soon finding a home for disabled people which offered him room and board to teach English to the residents.

I went to so many towns by train…taking the long overnight trips as well as short trips during the day…as it was so inexpensive to travel there. An overnight trip was about $35 and town-to-town only about eighty cents. Riding in the four-person bunk rooms in the overnight trains were a unique experience for me. Randomly placed in a cramped four-person bunk with strangers…mostly whom I could not communicate with…was less awkward than I imagined. All manner of men and women of all ages just went about making their bed and going to sleep. No fear or awkwardness seemed apparent in anyone…and I constantly saw people just routinely helping other people as a matter of course. Passing by fields, the corn stalks stood in shocks and the hay was mounded up in haystacks…like it was here in the 40's. There were no lawnmowers…even in yards and parks in town, people cut the grass with a scythe and fed it to livestock.

My first train trip was an overnighter from Kiev to Lviv. I wandered around Lviv for a couple hours sightseeing before meeting up with the Appalachian-Carpathian Conference folks. Serendipitously, while walking around, I walked straight to the National Forestry University of Ukraine without knowing there even was one. The buildings and grounds were beautiful, and almost every tree was labeled with its Ukrainian and Latin name. A young man walked out of the main forestry building as I noticed the plaque on the wall and realized it was a Forestry University. He happened to speak English and told me a bit about the place, but had to hurry off to a lecture. The synchronicity that seemed to be happening continued for the entire trip and time after time I felt like I was "in the right place at the right time". I had learned a few phrases in Ukrainian, a few words in Russian, and could read their Cyrillic alphabet…thanks to meeting with a Ukrainian Berea College student for a few weeks. But, I had scarcely traveled abroad…not at all where few people spoke English…and was not even used to using public transportation…yet, as I extended my trip to include a week of exploration alone in this fascinating country, everything seemed to fall into place.

The Museums of Kiev filled my first couple of days, after arriving there at the airport. Riding in the taxi cab to my hotel near the old city center, I saw the many, many hulking collectivist Stalin-era apartment buildings. These crumbling, ugly buildings gave way as we entered the old, pre-Soviet city center and then all the buildings were utterly beautiful and in far better condition than the much newer Soviet buildings. Beautiful Orthodox churches were everywhere, and I marveled that they had not been destroyed or altered during Soviet times. Beautiful, huge and heroic-looking statures and monuments were everywhere too. It seemed that an immense statue of some kinds was on every block.

I had not planned my route, but rather checked in to a motel I picked to be within walking distance of the train station, and then wandered wherever inclinations directed me. Following the walking trails and parks along the Dnipro River, I walked up on a small group of people waiting in line to go into what looked like a little cellar in the ground. A sign next to the "cellar" showed that it was a museum of water. "Cool…Berea College is a managed watershed…perfect place for me to go." As soon as I got in line, the door opened to reveal stairs going underground…the museum was all underground. Everyone else had tickets and knew what was going on, and the ticket lady saw I didn't have one and asked "Exclusive? Exclusive?" It seemed like "Yes" was the right thing to say, so I answered yes in Ukrainian. She motioned for a young lady to come up the stairs, and she greeted me in English…"Exclusive means you get your own tour guide, me, in English!" and guided me through the extensive museum. The museum was based on a prestigious museum in Denmark and one of the best water museums in the world, I learned.

I ended up continuing to walk south along the river and there was one museum after another…for about $2 American money per museum, these were some of the best museums I have ever been to. I walked all the way down to, and went to the World War II museum and the nearby Motherland Monument…an immense 203 foot tall stainless steel statue of a woman holding, in raised hands, a sword in one hand and a shield in the other hand. On the shield is the state emblem of the Soviet Union. While sneakily walking along beside a tour group with an English-speaking guide for a short while, I learned that most of the communist monuments had been taken down, but this one was considered a World War II monument and allowed to stay. I just learned, while writing this book, that the Soviet emblem is to be removed.

The tour guide was very critical of the Soviet era, but also of the corruption in her country since then. Everyone I chanced to find to talk to in English was quick to complain about their government's corruption and economic struggles. In the West, where much of Ukraine had formerly been part of the Austro-Hungarian Empire, Poland, and Lithuania over the centuries, most people had hope their country…or at least their part of it…would align itself with the European Union and they were worried about Russian aggression and interference in their country. But, they also worried that all the rules and regulations that would come, along with European Union membership, this would ruin the small farms, farmer's markets and "mom and pop" restaurants. In the East, most people seemed to favor merging with Russia or at least aligning with Russia politically…blaming the United States for meddling in their elections and world affairs. Some just threw up their hands and said at least they would rather their own politicians screw them over than European, American, or Chinese stealing from them. "One politician sells us out to China, and the other one to the West…otherwise they just steal for themselves." This was said in response to wanton clear-cutting of timber that just kept happening somehow…and was resulting in widespread erosion and sedimentation problems in the rivers.

The last museum I went to was the Museum of the Holodomor. The museum tells of the horrific Stalinist genocide against the Ukrainian people… the "Great Famine" that was a result of forced collectivization of the farmers and confiscation of not only the food they produced…to be sold for money for armaments…but even the seed needed to plant the next year's crops. This man-made famine occurred in 1932-33 and reportedly killed 7 to 10 million people. It happened before World War II and one can understand how some Ukrainians initially viewed Germany as an ally in the struggle to free Ukraine from Stalinist oppression. Statues of Stepan Bandera…the Ukrainian political activist and nationalist in the independence movement who collaborated with the Nazis…are prominent in western Ukraine. We ate at the popular "Bunker" restaurant in Lviv, which has a nationalist them. To get in this rather hidden, subterranean place you have to shout out the patriotic slogan "Slava Ukrayini!" (Glory to Ukraine!)

I was fortunate to be invited to present my forest research at the 2017 International Appalachian/Carpathian Conference in Ukraine in September. Sponsored by the Ivano-Frankivsk College of the Precarpathian National University, Berea College, the Yaremche Center for Social and Business Initia-tives, the International Programs of the U.S. Forest Service, Fulbright Ukraine, the Carpathian National Nature Park, and distinguished Appalachian Studies scholars from the USA, the program featured sessions on topics ranging from sustainable community development, cultural preservation and folklore, regional history, sustainable agriculture and forestry, ecological tourism, rural place-mak-ing, and other issues related to mountain life and culture in the Appalachians and Carpathians. Berea's Chris Miller, Associate Director & Curator of the Loyal Jones Appalachian Center at Berea College, has been attending these conferences for some time, and helped me get acquainted with this beautiful country and its people.

The conference itself was really interesting and educational, with profession-als from the USA and Ukraine presenting on a wide range of topics that all related in one way or another to many of the environmental and societal issues both of our mountain regions contend with. However, just getting around in Ukraine and experiencing the culture and conditions in this former Soviet nation was really enlightening. Staying at a hostel several nights (for $9 a night) and interacting with the people there…from Ukraine, Russia, Turkey, Norway and other countries…and hearing their perspectives, was remarkable.

Ukraine is, by American standards, a "poor country". Large numbers of people must travel outside the country to find employment. I heard of a medical doctor who has to pick grapes in Italy seasonally to heat her home through the winter, for example. A much smaller percent of the people own personal vehicles, and large numbers of people rely on family gardens to have enough to eat. It was the fall mushroom season, and it looked like nearly everyone was into looking for and selling mushrooms.

I have to say as a forester that the highlight of the trip was my excursion to the Ancient and Primeval Beech Forests of the Carpathians. The Carpathian Biosphere Reserve is another UNESCO World Heritage Site; the largest in Europe, in fact. Here, the largest and oldest stands of European beech (Fagus sylvatica) can be seen, and I was able to tour it with two Ukrainian foresters; one who has worked at this same location for 40 years. He smiled the whole time, and seemed as excited to show it to me as the first time he had given a tour… and we had a difficult time keeping up with him, too.   He said only about 20 people got to see this place of the forest per year; and mostly these people were German or Swiss researchers who came to re-measure research plots on the forest. Old signs from the Soviet era were still located in the forest, and he explained what they said and meant…they were forest community types. Evidently, the Soviets had quite an extensive interest in site-typing the forests and keeping up with them.

We found quite a few of the big white bolete mushrooms, and the forester picked the largest one and we brought it back to his office for the three of us to eat together. It was that big. His lovely wife sautéed it for us in a skillet, and we had it with cognac…delicious!

One particular little meadow within the Primeval Beech Forest was called "Gypsy meadow", and the forester explained this was because gypsies once roamed through there and would use this meadow to graze their animals. It seemed incredible that such a pristine area had been used by humans for hundreds of years and yet no one had cut down this stretch of forest. I asked him what happened to the gypsies, and he became quiet, then slowly told of how they were rounded up during the Nazi occupation and sent away…perhaps to their deaths. My simple question, thus, elicited a sickening feeling that was in sharp contrast to the festive mood we had been in. I chose to mention this in

my book because…while I celebrate German forestry and culture in my book…
I feel that it would be incomplete to not touch on this part of the history as well.

So, Ukraine was a land of contrasts for me. If there was ever a place
where…as I have heard "old timers" say of living through the Great Depression…where hard times brought out the best in people and caused them to work
together and help each other to get by; it is Ukraine. Young people still knew
how to do traditional farming practices and cook from scratch…yet most had
cell-phones and were as up-to-date with modern technology as anyone in the
USA. In some ways, yes, it is a poor country…but in many ways they seemed
to be the richest of people to me.

I am not well travelled, but these two trips; one to Germany and a year later
to Ukraine, were both formative experiences. I am excited to be accepted again
to speak at this year's Appalachian-Carpathian Conference and get to travel this
time to Romania. My presentation will be on sustainable forest management
using horses. Animal power still persists across much of Eastern Europe's rural
areas. I hope my presentation encourages people there to continue using animal
power as much as possible in their forests and on their farms.

If I had the money…or the courage that the old Irishman who hitchhiked
to Ukraine with $30 in his pocket seven years ago did…I would wander the
world. I hope this book helps open doors that will allow me to make more trips
like this to visit, study and write about sustainable forestry efforts in other parts
of the world. Travel…international travel in particular…is invaluable to understanding what possibilities and perspectives are out there to learn from and
adopt for your own or modify. While I definitely value scientific research and
do not want to downplay the role science plays in forestry…at the same time, I
see that exploring…in this case forests…and interacting with people who live in
and around them, depend on them, love them…can be so useful in learning
better practices by seeing them in action. It is the interaction of people with
their places, the nature and customs, traditions, and culture that makes a place
what it is.

Forced collectivism in the East and corporate hegemony in the West seem to
both homogenize and deconstruct the ecosystems and ecosystem-like human
cultures they subsume, and create monocultures in their place. This reminds me
of the characterization of capitalism and communism as "merging traffic" by
Edward Abbey, anarchist author of Desert Solitaire and The Monkey Wrench
Gang. The health of forests and the art of forestry are both casualties in these
two systems. I am so very thankful that my journey to the forests and foresters
of Germany allowed an encounter with another approach where both can thrive.

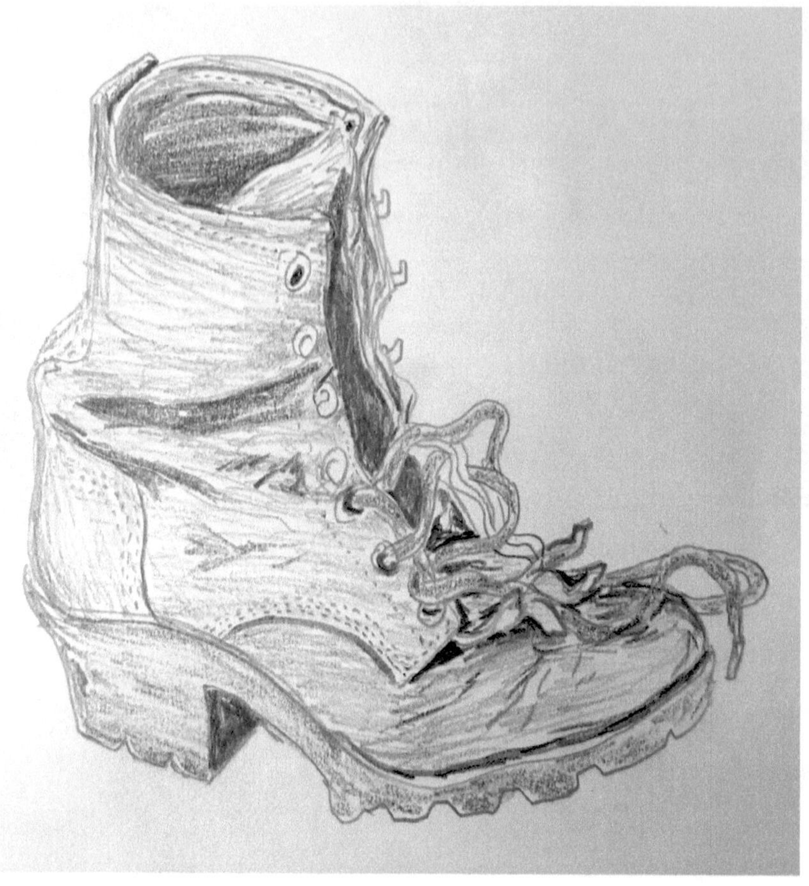

Chapter 6:
Work

"The mountains are calling, and I must go, and I will work on as I can, studying incessantly."
 – John Muir, in a letter written from Yosemite Valley, September 3rd, 1873.

I have a t-shirt with John Muir's famous quote: "The mountains are calling, and I must go." But, that is the abbreviated version. I have never seen the whole quote on a t-shirt. I suppose it is more romantic, or just more palatable, for us to think of Muir as just going off to the mountains to hike and enjoy nature casually. Perhaps he would write some inspiring poem or article for one of the magazines he was writing for. But, Muir was actually working for Yosemite…and studying incessantly!

According to Zoologist and Muir contemporary, Henry Osborn: *"Daily he rose at 4:30 o'clock, and after a simple cup of coffee <he would write> incessantly…"* Yes, Muir was practically a "workaholic", constantly learning, and a prolific writer too. http://www.thebasinandrange.com/get-know-story-behind-muir-mountains-calling-quote/

Yes, Muir was a "workaholic"; constantly learning…and a prolific writer too. I believe that for a "land ethic" to succeed, it must be paired with a strong work ethic. Aldo Leopold certainly applied a strong work ethic throughout his life, too. I have toured the Leopold "Shack" near Baraboo, Wisconsin and the Aldo Leopold Center nearby and see how he and his family "practiced what they preached", together, on their sandy hill farm. They were incessantly planting trees and tending to all manner of work on the land; as well as studying and reading all the time.

Physical labor is healthy too…for both the body and the mind. Exercise prompts the release of endorphins; which makes physical activity anti-depressive. As one of the least healthy demographic areas in the United States, Appalachia needs this. Yet, just about everywhere seems to be undergoing an increase in obesity, depression and an overall sense of a lack of hope and purpose. The forests need us and we need the forests. So, what better way to improve them both than working in forests to make us both more healthy and resilient? This kind of activity would promote a sense of camaraderie (community) and help engender a land ethic. Thus combined, we achieve community forestry.

Wouldn't it be wonderful if a strong work ethic and a strong "land ethic" were regularly employed together? What if, when the "mountains are calling", we "worked like the dickens" for them and also were constantly studying so we can "think like a mountain" and act out on our ideas? Then, we can share these ideas and accomplishments too; through writing and creating tangible examples for people to see!?

———————

Brownstown, IL (around 1980) "Run and get the spade, and lets go hunt some trees to plant!" Grandma hollered out the window. In a flash, we were "barreling" down the road in the "Old Blue Goose"; Grandma's 1950-something baby-blue car with the goose hood emblem. She'd drive fast over the little hills to make your belly flutter. Of course, back then kids stood up in the seat, and the car didn't even have seatbelts. Soon, we arrived at an old country cemetery bordered in the back by woods. "Here, stick that brick behind the tire before I let off the brake", she said, pointing to a brick in the floorboards, "That'll keep it from rolling down the hill." The "Old Blue Goose" had been known to do that.

Traipsing across the cemetery, we came to the woods and started picking out trees to dig up. "Shoot, there's way more trees here than there is room for", she said as she began to dig up a five foot tall sugar maple. "Hooey! That ground's hard. I can't take it with these bare feet", she exclaimed as she handed me the shovel. It was a cool fall day, but Grandma never wore shoes if she could help it. Jumping up and down on the spade, it took all fifty pounds of me to get it in the ground far enough for us to dig the tree out.

We had three sugar maples and a redbud dug, before we realized how heavy they were. "Here, you wait here, I've got an idea", she said, as she headed towards the car. Minutes later, she arrived with the spare tire cover and an extension cord. One at a time, we rolled the root ball onto the tire cover, tied it up with the extension cord, and dragged it across the cemetery to the car. Each time we'd stop to rest along the way, she'd point out a gravestone of someone she had known and talk about them. "Shoot, I guess I've got more friends in here than living… might as well get something done while I'm still here, that's what I say", she chuckled as we'd begin dragging the tree again. "There's nothing some know-how and elbow grease can't do", she stated, as we managed to heave the last one into the trunk.

By supper time, the trees were planted. Over the next three years, several more were planted by the same method by various grandkids. Remarkably, every tree lived. Some had to be taken out as they became too overcrowded. I think Grandma kept planting them, even when there were already enough, just so every grandkid got to do it. Today, these trees are approximately fifty feet tall and are a living reminder of Grandma and her remarkable "gumption", as she would say.

Four fairly large sugar maple trees were located in Grandma's yard which had been planted by her and several of her seven children before my time. One of these trees; with branches low to the ground and spaced apart just right, was referred to as "the climbing tree" by us grandkids. We spent so much time in that tree that each of its limbs we sat on had a certain name. Whether we were pretending to be Tarzan, monkeys or pirates, this was the place to be.

These tree planting expeditions are my earliest memories of planting trees, and probably influenced my "when there is a will, there is a way" attitude about getting things done, too. Grandma Torbeck used to tell us grandkids when we were having trouble doing something…or if we complained about having to do it…that all it took was "a little know-how and enough elbow grease" to do anything. I wasn't too sure what elbow grease was; only that I was supposed to use it. "Sweat? Does she mean sweat? "I thought the first time she said it, as I did have sweat dripping off my elbows by the time we were done with these projects that started out with those words.

When she put out a garden with just a spade and hoe, or mowed her lawn with a push mower (even after she needed a walker to get to it), it took a lot of "elbow grease". When Grandpa refinished all the woodwork in their house by scraping off seventy years' worth of varnish and paint with pieces of broken glass salvaged from a hail storm, it took a lot of it. It seemed like there was never a reason not to dive into any project that "needed done"; because, as they would always say: "All it takes is a little know-how and plenty of elbow grease".

Since then, I've noticed that most people have come up with substitutes for elbow grease. Leaf blowers, for instance…when you just have a little yard. Why in the world would anyone pay a bunch of money for one of those things, to stand around with that awful noise blaring, when all it takes is a rake? Well, I use them in big areas or on fire lanes, but for a tiny yard? Mechanical hedge clippers, too… and riding mowers for small yards in town… electric pencil sharpeners… dish washers… what people won't buy to get out of using a little elbow grease.

All these noisy, expensive, breakdown-prone contraptions... do these really make us better off? One thing they do make us is more dependent on petroleum. Petroleum might be said to "make the world go around", but before machines and money (or, more commonly, credit) became the go-to for getting things done, "a little know-how and plenty of elbow grease" is what we made do with. People tended to be healthier and in better shape, too.

For those who may not have heard of the term "elbow grease", it basically just means hard work. According to the website www.prhases.uk, the meaning is "energetic labor". The term has been around since the late 1600's, and its inclusion in a dictionary in 1699 suggests that it was a lower-class term. That makes sense, as the upper class probably didn't have to work anyway.

It sure seems like, sooner or later, petroleum and/or money are going to run out. We'll have to once again go back to depending on elbow grease…but where's the "know how" going to come from? The old cultural practices need to be kept alive. All the "old folks" I knew who grew up in the Great Depression, like my grandparents, talked about how they really liked it back then. Of course, these were hard times, but people helped each other and worked together.

Elbow grease was constructively used for programs, too. The Civilian Conservation Corps, for instance, provided Grandpa with a job when he turned eighteen. Sure, it took funding, but so many young men learned skills and a work ethic which benefited our country through the subsequent world war and through the rest of their lives. He was sent to Rhinelander, Wisconsin, to build bridges, pavilions, and such in State and National Forests. The work of the CCC's still stands in National Forests and Parks across the country and is a testament to "know-how and elbow grease".

I remember Grandpa telling me about how he came to appreciate what America was about while in the CCCs. He worked alongside men from all different nationalities, races, and backgrounds from across America. He described how in most American cities, the major churches are built clustered around the downtown (before so many built new mega-churches out by the interstate highways, like big-box stores, instead) with many different denominations right beside each other…with no fighting and strife. Workers of all different backgrounds, races and religions pitched in, as equals. This is what it is to be an American. He told me that he believed the garbage man should not make less than half what the President of the United States makes, for example, and that every job is important and all workers should be treated with respect.

Berea College, as one of America's handful of "work colleges", is a place where students are engaged in applying some "know how" and "elbow grease". According to Berea College's website:

The Student Labor Program originated in its earliest form at Berea College in 1859 and expanded to become one of the College's Great Commitments. The Labor Program provides economic, educational, social, personal, and spiritual benefits to students and those served by their work.

The Labor Program is designed to serve the following purposes: support the total educational program at Berea College through experiences providing the learning of skills, responsibility, habits, attitudes, and processes associated with work; provide and encourage opportunities for students to pay costs of board (meals), room, and related educational expenses; provide staff for College operations; provide opportunities for service to the community and others through labor; and establish a lifestyle of doing and thinking, action and reflection, and service and learning that carries on beyond the college years.

Designed to serve these multiple purposes, the program reflects a unified vision of labor as student and learning centered, as service to the College and broader community, and as necessary work well done. The administration of the program is the responsibility of the Dean of Labor.

Labor assignments function very much like classes. Beginning at entry levels of work, students are expected to progress to more skilled and responsible levels. Through these experiences, it is expected that student workers will: develop good work habits and attitudes; gain an understanding of personal interests, skills, and limitations; and exercise creativity, problem-solving, and responsibility. Students also may learn the qualities of leadership, standard setting, and effective supervision.

The Labor Program makes it possible for students to know each other as co-workers as well as classmates. More importantly, linking the Academic and Labor programs establishes a pattern of learning through work that continues long after college. (http://catalog.berea.edu/en/Current/Catalog/Admissions-and-Financial-Aid/Copy-of-The-Labor-Program)

It has been my experience that much of the most important learning happens in life application and work. This applies not only to learning how to apply technical information, but the "people skills" that are so important to many fields in today's workplace. Perhaps more important even than workplace skills, however, is the kinship and mutual respect that it engenders in people.

Dr. Martin Luther King, Jr. delivered his "I've Been to the Mountaintop" speech in support of the striking sanitation workers…garbage men… at Mason Temple in Memphis, TN on April 3, 1968. I traveled on the Berea College Civil Rights Tour to the site in Memphis where he was killed. It sounded to me like, from the tour information; that Dr. Martin Luther King's support for the common worker was beginning to resonate with people of all races. The common working man and woman…regardless of color…was beginning to hear his message and think "Amen…me too".

I wish the United States still has something like the Civilian Conservation Corps today. It seems even more important now than ever…perhaps most of all in our young men…who are growing up far too often without a father or good father figure to teach them to be respectful toward women, how to fix things, and the importance of hard work and responsibility. I respect those who choose to go into the military, and many do find a sense of belonging, service, and career training in military service to their country. However, it seems that there should be something else available as well…for both men and women.

Yes, the Dignity of Labor, the Kinship of All People, and Gender Equality all do go hand in hand, in my experience. People who might otherwise be fighting with one another…if joined together in a common cause…such as work…can become friends. Berea College's stance that we are all "created of one blood" and its history as a work college go hand in hand. Martin Luther King, Jr., in his speech to the sanitation workers referred to us all as "God's children", and sought fair treatment for everyone. Working together, in kinship and with dignity of labor, we can overcome our societal struggles. According to Martin Luther King, this can only be achieved without violence… *"It is no longer a choice between violence and nonviolence in this world; its nonviolence or nonexistence."*

Much can be accomplished when people work together. "Many hands make light work" as they say…while "idleness is the devil's workshop". Sometimes, I think our culture elevates competition too much; to the detriment of cooperation. In some ways, I think the "survival of the fittest" mantra contributes to this idea. Yet, when I look at Nature, I do not see a "free for all", I see networks of cooperation everywhere. The very existence of an ecosystem belies it, and scientists are beginning to realize and accept that this cooperation much more highly communicative than ever before conceived. Peter Wohlleben, in his hit book The Hidden Life of Trees: What They Feel, How They Communicate – Discoveries from a Secret World eloquently introduced this concept with trees:

"When trees grow together, nutrients and water can be optimally divided among them all so that each tree can grow into the best tree it can be. If you "help" individual trees by getting rid of their supposed competition, the remaining trees are bereft. They send messages out to their neighbors in vain, because nothing remains but stumps. Every tree now muddles along on its own, giving rise to great differences in productivity. Some individuals photosynthesize like mad until sugar positively bubbles along their trunk. As a result, they are fit and grow better, but they aren't particularly long-lived. This is because a tree can be only as strong as the forest that surrounds it. And there are now a lot of losers in the forest. Weaker members, who would once have been supported by the stronger ones, suddenly fall behind. Whether the reason for their decline is their location and lack of nutrients, a passing malaise, or genetic makeup, they now fall prey to insects and fungi.

But isn't that how evolution works? you ask. The survival of the fittest? Their well-being depends on their community, and when the supposedly feeble trees disappear, the others lose as well. When that happens, the forest is no longer a single closed unit. Hot sun and swirling winds can now penetrate to the forest floor and disrupt the moist, cool climate. Even strong trees get sick a lot over the course of their lives. When this happens, they depend on their weaker neighbors for support. If they are no longer there, then all it takes is what would once have been a harmless insect attack to seal the fate even of giants."

– Peter Wohlleben, The Hidden Life of Trees: What They Feel, How They Communicate – Discoveries from a Secret World

---

"Survival of the fittest?" This thinking reminds me of something Gandhi was attributed as saying: "An eye for an eye will leave everyone blind". Yet, Nature does not appear to be blind to me. A mere "survival of the fittest" system would, it seems to me, not have resulted in the vast diversity of life on this planet. Only with Man as "ultimate predator" acting out in a "survival of the fittest" mentality, do we see a collapse of diversity and Homo sapiens proliferate at the expense of all others. I think we are beginning to realize that ecosystems function through everything working together; not working against each other. It may be that "survival of the fitting-in-est is more accurate.

These ecosystems developed over thousands of years. When we humans directly destroy habitat, it causes the largest negative impact. But, as we rapidly disperse species from around the world into different ecosystems, this also upsets the cooperative system where everything that was there before fit in. Invasive species are now one of the biggest threats to ecosystems. In the Berea College Forest, for example, bush honeysuckles *(Lonicera maackii, L. tatarica, L. morrowii, L. x bella)*, Chinese privet *(Ligustrum sinense)*, Tree of Heaven *(Ailanthus altissima)*, Oriental bittersweet *(Celastrus orbiculatus)*, and kudzu *(Pueraria montana var. lobata)* are our major invasive plant problems. All of these species come from Asia. We spend a lot of time cutting them, applying herbicide to them, pulling them, and recently have begun using goats in an attempt to control the kudzu.

Invasive pests such as the emerald ash borer *(Agrilus planipennis)* is wiping out the ash across eastern North America, while the woolly hemlock adelgid *(Adelges tsugae)* is killing off the hemlock trees throughout their native range. We have treated our hemlock trees…which are important in creating a shaded micro-environment along cool, spring-fed streams…with imidacloprid insecticide. So far, the treatments have been successful in saving the trees, but the feasibility of repeatedly treating them if the adelgid threat remains, is questionable. Treatment of ash, with the same insecticide, was less successful…only about one-third of the trees treated survived emerald ash borer infestation.

Eventually, perhaps new cooperative systems will become established where this artificially imposed "survival of the fittest" scenario was injected into the existing, cooperative native ecosystem. New cooperative arrangements are nothing new, but the rapidity of introduction of outside species into these systems is perilous and detrimental. We are still reeling from the loss of the highly important American chestnut *(Castanea dentata)* tree from chestnut blight…caused by the fungus pathogen *Cryphonectira parasistica*…also from Asia.

Species keep getting decimated; not quite made extinct, but knocked into a state of barely surviving, functionally extinct and no longer an important component of the ecosystem. The cooperative ecosystems are disrupted…much faster than species can recover. The fragmentation of the landscape exacerbates this problem; isolating genetic information. New kinds of forest management strategies need to be devised and employed to help the ecosystems recover. Foresters can play a key role in developing these new strategies.

———————

"Resilience" is a relatively new "buzzword" that I have noticed has emerged over the last fifteen years or so in forest management circles. Forestry professors are starting to talk and write about "forest complexity" and "forest structure" as being important…not just wood volumes and growth rates. I think the profession is finally splitting into two branches, as Ado Leopold advocated: agronomic forestry for plantations, and ecological forestry for natural forests. This is a good thing in my estimation…as we need plantations to produce as much as possible so we don't have to use up our native forests as heavily to supply wood fiber. And we need our native forests to continue to be native, intact ecosystems. A different mindset is needed for each branch. The agronomic model is "going gangbusters" already…genetic modification and all sorts of technological applications and capital are employed in this arena. However, the ecological branch is new.

Managing for forest complexity…rather than homogenization…is becoming an important component in efforts to increase or maintain forest resilience. Forest resilience is the term used to describe the forest ecosystem's ability to maintain itself amid the ever present and most likely increasing onslaught of outside change agents. Introduction of invasive species is a big change agent. The removal of fire from the landscape of North America has been a big change agent, as has the resultant catastrophic-level hot fires in some parts of the continent since fuels have accumulated from fire removal for so long. Perhaps the change agent that gets the most attention, is climate change. Whatever change agents are acting upon our forest ecosystems, their resilience in the face of change is highly important to the maintenance of these ecosystems.

I decided to go back to college, part time, to earn my master's degree in forestry at the University of Kentucky to be a part of the solution to help find more effective, more "natural" ways for managing our native forests to help them to be more resilient ecosystems. Dr. John Lhotka, also an alumni from Southern Illinois University at Carbondale…had an interest in researching gap-based regeneration methods for mature upland oak forests, and had been involved with research projects on the Berea College Forest along with Dr. Jeff Stringer, University of Kentucky Extension Forester, and long-time expert in forest management. They had been researching edge effects and midstory removal effects on oak regeneration at Berea College already, so my timing was perfect to get involved with their efforts as a graduate student while working as Berea College Forester.

I have to admit, I was slow to "warm up" to the whole concept of research. Primarily, this is because I could not wrap my head around the logic of purposely doing something randomly. Also, I wanted to track down the outliers and study them…not throw them out as lying too far outside what was the "normal" distribution. I wanted to put my harvest gaps where there was advanced regeneration (tree saplings already established enough to compete successfully) of oaks because that's where I thought it would be successful. I wanted to make the gaps the size and shapes I noticed natural canopy gaps producing the advanced regeneration were. Not allowed.

Goethe's approach to science, from what I had read of it, seemed to make more sense to me. I felt like the many hours I had to spend in front of a computer entering data and trying to analyze it statistically, could not possibly reveal more to me than sitting on a log just observing a natural canopy gap that was successfully recruiting oak. At best, it seemed like I was producing only a tiny artificial key-hole-sized glimpse of what we hoped to see…while ignoring what was promising results that were naturally happening nearby. I imagined myself being forced to look through either a microscope or a telescope with one eye to try to see the "big picture", when depth perception takes both eyes. I like to think of science serving the role of one eye, and the arts/spirituality as serving the role of the other eye. By the time I was done, I felt like I had been jerked through a keyhole myself.

If only a master of arts was available in forestry! I could go for that…maybe even a doctorate degree. I'd love to that…but no more of this tedious, analytical stuff. Still, I am so grateful to have had the opportunity to interact with the University of Kentucky forestry professors. My research project is titled *Initial understory response to gap-based regeneration methods for mature upland oak forests*. I present here a summary of the research project; along with the pertinent literature cited in it, to explain the broad challenges that forest management faces in our oak forests today and to give credit to the researchers who have dedicated their lives to helping find viable solutions to these challenges.

Gap-based regeneration methods as a forest management practice in the Central Hardwood Forest Region (CHFR) are gaining importance with the recent interest in longer rotations and in reproduction methods that retain an overstory for extended periods. My research project examines the two-year initial understory response to a gap-based regeneration method and associated midstory removal. The stated intention of the study is to provide information to help guide forest management decisions in the region such that forests can be managed profitably while promoting the regeneration of oak and maintaining aesthetics. More broadly, however, is the larger goal of maintaining the oak forest as a whole, not simply as stands.

Tree species composition in eastern hardwood forests, in the absence of historic fire regimes, which promoted oak species, is trending toward shade tolerant species in the understory (Smith 2005). Trends toward smaller forest tract size are also leading to an increasing aversion by landowners to large visual impacts (Smith 2005). As the shade tolerant species eventually replace the mature oaks in our forests, serious economic and environmental impacts loom. In Kentucky, where 48 percent of the state is forested, 75% of this forest is oak-hickory, 299 million cubic feet of wood is harvested annually, and 88.5% of the forests are privately owned (USFS & KDF 2008) this issue is particularly important. Forest management objectives across the CHFR increasingly focus on oak *(Quercus)* ecosystem management priorities for both timber and non-timber objectives that rely on maintaining oak-dominated stands to promote understory plant diversity, game wildlife productivity, and neotropical songbird habitat (Groninger 2008). Additionally, invasive exotic species (many of which are promoted by disturbance), increasingly invade and adversely impact our forest lands (Aukema 2010). A harvest method that does not create such widespread disturbance, maintains understory plant diversity and aesthetics, but also promotes oak regeneration, is needed.

The project refers to the areas being harvested as "gaps". Research in the area of gap-based silvicultural systems sometimes refers to harvest openings of various sizes by different names according to gap size, placement, and purpose. Terms such as "group openings", or "patch cuts", "single-tree gaps", "regular shelter-wood" and "irregular shelterwood" may all be used to describe gap-based silviculture systems (Kern et al. 2016). An edge environment with greater light availability for seedlings created around gaps has been shown to extent at least 10-30 m into the surrounding forest matrix (Gehlhausen et al. 2000; Matlack 1993; Tryon et al. 1992; Hamberg et al. 2009). *Femelschlag* systems utilize an expanding-gap irregular shelterwood to gradually expand harvest gaps into this edge environment; thus providing a continuous process of specific species regeneration around enlarging gaps until the forest is regenerated (Spurr 1956; Raymond et al. 2009).

The goal of the project is to develop gap-based silvicultural practices that address the oak regeneration problem present in mature oak dominated stands within the CHFR while providing a sustainable management approach that promotes forest health, structural complexity, and species diversity. Many variables influence regeneration dynamics in forest gaps; such as gap size, canopy and midstory structure and edge effects. Understanding these variables and their interaction is an important step towards developing gap-based methods that can be utilized in mature oak forests. These methods are commonly used in Europe, but are not widely documented in North American oak forests. This study is designed to address the lack of research in the use of such gap-based harvest systems for mature oak forests. Light transmittance and two-year seedling response data following a harvest utilizing 60m diameter gaps and associated midstory removal around these gaps provided the basis for this project.

The study examined how oak and its major competitors yellow poplar and red maple interact in the understory following a 60m (200 ft.) diameter gap harvest with and without midstory removal in a 30m band surrounding the gap after two growing seasons. Specific objectives are described in three categories below, but all are aimed at better understanding the interactions between the gap harvest, midstory removal, and the growth and development of seedlings exposed to this combined treatment.

Uneven-aged methods like group selection, even-aged methods which utilize patch cuts, and the irregular group shelterwood *(Femelschlag)* system of Germany and Switzerland, are all part of the framework being considered in the development of a more unified gap-based system for the Central Hardwood Forest Region.

Regenerating *Quercus* spp. on productive sites in eastern hardwood forests is challenge to forest managers (Merritt 1979). The lack of oak regeneration in oak stands after various methods of harvesting is a problem across much of eastern North America (Lorimer 1983, Smith 1993). This is of particular concern, as oaks are the most valuable hardwood species group in the eastern United States (Luppold and Bumgardner 2008). In addition to the economic importance of oak species, their value to wildlife cannot be overstated, as they provide the primary source of hard mast throughout their range (Goodrum et al., 1971; Pekins and Mautz, 1987).

Clearcutting and shelterwood regeneration methods, though extensively adopted and studied, have given mixed results (Loftis 1990, Van Sambeek 2003). Meant to promote oak regeneration, these even-aged methods are often successful on poor quality, xeric sites where advanced oak reproduction is present in high numbers but on intermediate to high quality sites, they are not reliable (Johnson et al., 2002). The light intensity needed for oak to reproduce adequately from seed has been considered to be at least 35 percent of full sunlight (Minckler, 1965). Another source indicated that oak seedlings (including red and white oak group) reach their maximum photosynthetic rate at about 30% of full light and this level was difficult to obtain through overstory manipulation alone because of pervasive understory competition (Sander, 1979). Even-aged management on higher quality sites tends to produce mostly shade intolerant species such as yellow poplar *(Liriodendron tulipifera* L.*)* (Kolb et al, 1990). Also, as the size of forestland tracts continues to decrease across the eastern United States, landowners are less willing to employ these harvest methods (Smith 2005).

Single tree and small group selection; which are more accepted by small private landowners and the public (Smith 2005), have been shown to convert oak dominated stands to shade tolerant species such as red maple *(Acer rubrum* L.*)*, sugar maple *(Acer saccharum* L.*)* and American beech *(Gagus grandifolia* Ehrh.*)* (Dey et al. 2008, Jenkins and Parker 1998, Lorimer 1983). A study in Indiana suggested that canopy gaps do not significantly benefit subcanopy populations of oak, due to insufficient understory light levels (Cowell et al. 2009). Likewise, harvests which concentrate on the use of group openings to promote shade intolerant species have shown that small openings will result in regeneration of the stand to shade tolerant species (Walters and Nyland 1989). Meanwhile, the "super-generalist" red maple, able to thrive across a wide range of moisture and light availability, is increasingly becoming a main component of the understory in oak forests (Abrams, 1998). To maintain oak composition near historic levels, silvicultural strategies are needed that will result in an accumulation of advance oak regeneration instead of red maple and other shade tolerant species, before the canopy trees are replaced (Johnson *et. al.* 2002).

Gap-based regeneration methods for mature upland oak forests could prove to be a beneficial management practice for regenerating oak; particularly where the maintenance of an overstory for an extended period is desired. Sometimes considered a "nature based" silviculture, (where managers try to emulate natural disturbance through creating small openings which approximate windthrow, natural mortality, etc.), gap dynamics has been used to successfully assist in regeneration (Emborg 1997). Harvest methods using gaps of various sizes can be a good fit for aesthetics, wildlife and water quality (LeDoux 1999; LeDoux et al. 1993) by retaining high canopy and large trees on much of the landscape. Property value on small tracts may also be maintained through using gap-based harvest methods by not diminishing aesthetics (Atwood et al. 2009; Miller 1993). Use of gap-based harvest methods, when properly implemented, may improve timber quality and promote intermediate shade tolerant species such as oaks and hickories (Atwood et al. 2009; Ward et al. 2005).

The gap-bases system studied is commonly called an irregular group shelterwood *(Femelschlag)* and is a traditional reproduction method that has been used in Germany and nearby Switzerland to address these issues in mixed-species mountain forests with silver fir (*Abies alba* Mill.) (Puettmann et al. 2008). While early forest management efforts developed by Cotta and Hartig in 1804 and others advocated even-aged, homogenous stands (Kovac 2016), methods developed by Gayer in 1898 and others promoted irregularly-structured mixed forest stands and introduced irregular shelterwood and close to nature practices (Kovac 2016). Under the *Femelschlag* system, trees are selected for harvest based on spatial and stand structural considerations; opening up gaps, removing trees for poor health reasons, etc., and utilizing flexible and variable harvest schedules; with the underlying approach to management focused on a small spatial scale rather than a stand-level scale (Puettmann et. al. 2008). In contrast to even-aged systems developed for intensifying wood production, irregular shelterwood practices are grounded on the principles of spatial order, regeneration and tending of forest stands (Diaci 2006) to provide addition ecosystem services such as water supply, retention capacity of soils, carbon storage, air filtering, erosion mitigation, recreation, and other attributes (Kovac 2016). Such an approach may prove to be useful across other forest types.

In North America, irregular shelterwood methods may be effective in regenerating species such as oak, as well as the maintenance and restoration of irregular stand structure and other ecosystem-based objectives (Raymond et al. 2009). This study is designed to help understand how *Femelschlag* principles could be employed in the CHFR to develop specific strategies for targeted outcomes in oak forests.

My research paired a gap-based harvest with midstory removal inside and around the perimeter of the gap created by the harvest, and also included planting white oak seedlings. Midstory removal creates a wide array of light and microclimate conditions that affect seedling establishment, growth and competitiveness among various tree species. While the initial response of oak seedlings after two growing seasons did not yet reveal profound results at the statistical level, various trends do emerge. The increase of available light into the forest from the gap edge; particularly when coupled with midstory treatment, suggests that an enhancement effect for oak has been created based on position relative to distance from gap center and edge. A better picture of exactly where the optimum positions within and around the harvest gaps for oak regeneration exist, may better reveal itself in subsequent years.

The study lays the groundwork to help determine if gap-based regeneration methods for mature upland oak forests could prove to be a beneficial management practice for regenerating oak; particularly where the maintenance of an overstory for an extended period is desired. *Femelschlag* principles utilized in Central Europe, could be further studied in the Central Hardwood Region to develop specific strategies for targeted outcomes. In these systems, trees are selected for spatial and stand structural considerations; opening up gaps, removing trees for poor health reason, etc., and utilizing flexible and variable harvest schedules; with the underlying approach to management focused on a small spatial scale rather than a stand-level scale (Puettmann et. al. 2009, Röhrig et al. 2006).

Subsequent studies of these gaps is suggested, so that oak seedling response trends have had more time to become significant and reveal trends that could help develop a system for using a gap-based system in the CHFR. If such a system could be proven successful in regenerating oak, numerous benefits to owners could be realized. Potential advantages of expanding-gap systems include the maintenance of continuous forest cover and structural complexity, periodic and flexible income flows, regeneration of diverse species groups of varying shade tolerance, aesthetic maintenance of the view-shed, wildlife habitat diversity creation and maintenance, and the promotion of a stable local wood products economy.

I am beginning to set up, with the help of our Berea College students and other faculty and staff at Berea College, more research plots on the College Forest to monitor change over time. In particular, as we move into our third year with a prescribed burning program, I hope to be able to quantify the response of the forest to these efforts. I don't have to do things randomly anymore…I am sizing new gaps and placing them strategically. Hopefully, I can get someone else to do the statistical analysis. The University of Kentucky continues to conduct research on the Berea College Forest as well, and I expect many beneficial things to come from their dedication and expertise.

———————

Kentucky is known for its race horses and its bourbon. *Old Forester*, my favorite, is the oldest bottled bourbon in Kentucky. Originally sold in pharmacies as medicine, the sealed bottle insured that it was not adulterated. I first discovered *Old Forester* while attending college as a forestry major. I wondered…were there any "old foresters" in Kentucky way back in 1873 when the product was introduced? Who could they have been thinking of when they named it? No, there weren't any professional foresters in Kentucky then. Silas Mason, as far as I can tell, was Kentucky's first professional forester. The product was reportedly named after the physician, Dr. William Forrester, who endorsed its consumption. The spelling is thought to have been changed so as not to reference a particular individual.

I can't help but think it would be appropriate to feature Silas Mason on the bottle of *Old Forester* and use his example and message of sustainable multiple-use forestry to promote responsible forest management. Bourbon depends on clean water and white oak, and Mason was a pioneer in insuring a sustainable supply of both of these. Berea College's spring-fed municipal water supply is one of the cleanest in the nation because of Mason's early vision for the forest to be managed as a watershed, and our white oak is also some of the highest quality in the nation for the same reason.

Berea and Berea College have a long history of being "dry"; and old newspaper articles on the subject leave no doubt the founders of the college would have been aghast to think of their spring water and white oak being used to produce bourbon. However, I would be surprised if Silas was (in private) a teetotaler, as he was from Vermont and went to college in Kansas. Brown-Forman, the parent company that makes *Old Forester* is the only bourbon distillery that makes their own barrels…making it even more of a fit for their *Old Forester* brand to be used to help promote responsible care of our forested watersheds.

The bourbon industry is booming…and the growing demand for Kentucky white oak barrel-aged whiskey is driving up the demand for white oak. This concerns foresters. We were already trying to fight for the continuation of oak forests; but now white oak is being sought after more than ever for barrel staves. The University of Kentucky's Forestry Department is naturally at the forefront of this effort and has begun a "White Oak Initiative" to focus attention and resources on white oak management. Berea College is a participant in this initiative, in the area of research and providing an example of sustainable management of this valuable resource.

The "Bourbon Trail" has become a major tourist attraction in Kentucky as bourbon distilleries have opened themselves up for tours and promoted distillery-hopping bus trips. The historic Boone Tavern at Berea…always called a tavern but never serving alcohol…was an enigma to travelers from areas where alcohol sales are taken for granted. Finally, it has begun to offer alcoholic beverages with a meal and has even been included on some of the distillery tours as a destination. I have served as a speaker to these groups from time to time, and focused my discussion on the history of our forest and its example in water quality and sustainable forestry. I have toured most of the distilleries in Kentucky myself; as well as the Jack Daniels' distillery in Tennessee. Jack Daniel's is nostalgic to me too, as it is another "spirit of St. Louis". Its popularity surged after it won a gold medal at the St. Louis World's Fair in 1904. Just six years later, in 1910, Tennessee "went dry" and Jack Daniels fled to St. Louis to resume making his whiskey. Ten years after that, Prohibition was passed nationwide and those who continued to make "spirits" had to ship it out discreetly.

The Berea College Forest's first forest management plan was written just ten years before Prohibition was passed. The many springs, natural rock houses, and wooded ravines made it an ideal place for bootleggers. One of the chief tasks of the College Forest tenants was, besides fighting forest fires, was chasing out the "pesky moonshiners" from the forest. Local newspapers consistently berated people from imbibing in alcohol. Bootleggers didn't have time to age their "moonshine" in oak barrels to turn it into bourbon…they had to ship it out before they got caught. Rumor has it that many shipped theirs out of the Berea College Forest in varnish cans to a Cincinnati "varnish factory" where it was distributed to Speakeasies. I wonder if any of it ended up in the small speakeasy near where I grew up in Illinois…home of Al Capone.

The little white building along the National Road (highway 40) was left much as it had been when the owner's husband was still alive many years ago. She was a friend of my Grandma Torbeck. I painted the old speakeasy the summer after I graduated high school. Inside, was a stage along the back wall and seats still set up. I wondered if any of the old fiddle-players who came to play in Grandma's basement had played there. As I painted the weathered wooden clap boards over with a fresh coat of white paint, my mind was drawn to wonder about the goings-on in this building way back when. Covering every inch of the building with my back-and-forth brush strokes, and looking away from time to time across the flat farmland and patches of trees, I could hear Interstate 70 to the south and imagined how the old National Road used to be the main thoroughfare. My aunt, Connie, helped get it designated the Historic National Road, an All-American Road. Before the interstate highway came through, all the little towns through here were bustling. Known as the Cumberland Road, this was the first major improved highway in the United States. Beginning in Cumberland, Maryland, in 1811, it reached nearby Vandalia, Illinois, before money ran out in 1837. At the time, Vandalia was the state capital of Illinois. The road was soon connected to St. Louis.

I took a break from the painting from time to time to cool off inside the dusty old building and gulp down some water from my big cooler jug. An old Brownstown Bombers jacket hung in the corner along with some stacks of old newspapers. Grandma had told me Gwen's husband never came home from a run, and that she used to stand out by the road watching for him. She speculated that the mob had got him.

I climbed back on my ladder and resumed painting well into the evening… trying to finish the south side while it was cool. The door to the house suddenly swung open: "Roy's not showed up for supper again! Dad blame it!" Gwen hollered. "Get on in here and have a bite to eat young man. I even baked an apple pie! You ought to be plenty hungry from being out there all day."

The table was set for three people. After I finished my delicious meal, Gwen looked at the floor for a couple minutes, then slowly got up and picked up the other plate and scooped the food into the trash. She then put all the plates in the sink and placed a piece of pie in front of me and retrieved an old black and white photo of herself and her disappeared husband. Between taking bites of home-made, delicious pie, I stared at the photo sitting on the table in front of me. It was of a beautiful young woman in a flapper dress and a handsome young man in a dashing suit…Gwen and Roy.

After finishing my piece of pie, I commented how nice they looked together and looked up at her to hand it back. Her aged, wrinkled face was beaming with a bright but quivering smile…and her eyes ran with tears. She softly took the framed photo from my hand and hastily went into the bedroom to put it up. As she walked away, her voice trailed off as she told me I should get back to painting and she'd do the dishes.

The west end of the Cumberland Road is a long way from the north edge of the Cumberland Plateau where the Berea College Forest lies. However, the networks to supply alcohol went far and wide. Johnson City, TN, is my wife's home town, and was commonly referred to as "Little Chicago" in the Prohibition era as it served as a major Capone stronghold between Chicago and Miami. Whether traveling from Brownstown or Chicago to Johnson City, TN, Berea is on the way. The Berean gentleman supposedly behind the "varnish" operation left no heirs and donated his farm and home to the College. I lived in this home for three years. It is a stately two-story log house with the logs exposed on the inside. I never found any varnish cans, but I did notice a hidden compartment under the closet floor. When I moved into the house, a kindly old neighbor showed me some old photos of one of the stills that used to be near here. One of his relatives was in the photo. They took a photo before destroying the whiskey and the still.

Bourbon barrel demand is driving the white oak market right now. It is interesting to tour a cooperage operation where the barrels are made. Foresters naturally like to learn how anything is made from wood, and I think continuing education opportunities which allow for professional foresters to become more knowledgeable in how timber is harvested, lumber is sawn, barrels are made and seedling nurseries operated, etc. is highly beneficial. Seeing first-hand how things are done in other countries and cultures, I think, is another wonderful way to expand our knowledge.

Silas Mason, the first Berea College Forester; as was the common practice at the time in the United States, went to Germany to learn forestry. Indeed, forest management in the United States owes much of its development to German foresters such as Carl Schenck (Vanderbilt's forester after Pinchot), Carl Schurz (Secretary of Interior) and Bernhard Fernow (chief forester of the USDA). Even America's own Gifford Pinchot, often regarded as the "father of American conservation" was mentored and particularly influenced by his good friend, Dietrich Brandis; whom the British put in control over India's forestry. (http:// www.gmbookchest.com pdf/Forestry_Politics.pdf 9/16/13).

Appalachia, as much as anywhere, benefited from the forestry knowledge of German settlers. The vast majority of Appalachia was populated by settlers moving into the region via the Shenandoah Valley from Pennsylvania. The first of these white permanent settlers were Germans. They were skilled at building with wood; resulting in the iconic "log cabin". They also developed and built both the Conestoga wagons that transported settlers and the Jaeger long rifles (which became known as Pennsylvania or Kentucky rifles) used for hunting and defense. Adam Miller (Mueller) and Hans Jost Hite (Heydt) have variously been credited as being the "first Appalachian". Both came down from Pennsylvania. Hite was wealthy, and obtained vast parcels of land on which he brought down many fellow Germans as well as Scots-Irish from Pennsylvania to settle... starting a pattern that continued throughout Appalachia (Historical marker in Winchester, VA). Our recently hired Horse Program Forest Technician, coincidentally, is also named John Hite (Hans is a derivative of Johann, or John).

At first glance, it would seem that because of the German influence on American forestry, forest policy in the two countries would differ little. However, the United States and Germany actually practice a decidedly different style of forest policy. Policy formation in the two nations and the process by which it is derived and administered is the result of varied historic, cultural, and economic conditions experienced in each country.

"Scientific exploration and natural resource management occur in direct response to human need" (Puettman 2009). Germany's pioneering efforts in forest management were a response to the human need for wood. According to Heinrich Cotta, sometimes referred to as the "pioneer of forestry", "There would be... no forest science without deficiency in wood supplies. This science is only a child of necessity or need" (Cotta 1816, 27). Germany, like other European nations, experienced a shortage in wood supplies during the eighteenth century. However, unlike the maritime powers of Europe, Germany was not positioned to exert itself abroad to exploit forest resources elsewhere (such as the Americas) when timber shortages loomed. Thus; while Britain, Spain, France, etc., were colonizing the globe and taking other nations' resources, Germany had to look inward to learn how to manage its own forests sustainably (http://www.wood-report.de/seiten/knowledge.html).

Germany made its move to "take over" land for "Lebensraum" (living room) much later; resulting in World War II. Strangled by the far flung Allies and their access to resources from around the globe, Germany's Axis was defeated and confined again to the densely populated core of Europe…forced again to look inward for solutions. Before they were defeated, due to their lack of access to resources around the globe, they invented the diesel engine to run on vegetable oil, wood-burning vehicles, and synthetic oil and rubber before the war was over.

Then, in the 1970's, a disturbing phenomenon began to occur throughout Germany. The foliage of many trees began to appear sparse and die off. Named *Waldsterben* (forest death), acid rain was blamed. The Forest Preservation and Forestry Promotion Act soon passed in West Germany in 1975 to help forests recover from the mysterious *Waldsterben*. Now reunified with East Germany; the Act now applies to all of Germany, and requires forest owners to return cut areas to their original condition (http://countrystudies.us/germany/144.htm, 9/17/13).

Facing the loss of their beloved forests, Germany became a global leader in "green technologies" such as solar and wind power, forest management, global forest certification efforts and climate policy (http://forestpolicyresearch. com/2009/01/25/germany-history-of-their-fascination-and-love-of-forests-and-trees/ 9/17/13). German citizens have generally been willing to go along with increased regulation and higher taxes to accomplish real, tangible improvements in the environment in general, and forests in particular.

Germany's geographic confinement to central Europe is a primary factor which explains many policy differences between the United States; which is much younger, less densely populated, and still had a resource-rich frontier fresh for exploitation little more than a century ago. Germany experienced wood shortages as a growing, industrializing nation with no colonies to extract from. Forest management sprung from this necessity. Cotta's premise, that "… no forest science without deficiency in wood supplies" thus explains much of the difference in forest policy between the two nations. Perhaps it could be said that Germany is simply ahead, so far as forest policy is concerned, on the continuum of time with regard to having to face the fact that resources are finite. While both World Wars are often attributed to Germany's aggression; their emergence as an industrial juggernaut came only after Great Britain had invaded and extracted resources from almost the entire planet already. Goethe, who lived during this time, lamented that while Germany was busy with philosophy, Britain was running the world with their practicality.

Besides geographic confinement, there are cultural differences between Germany and the United States which also leads to a different style of forestry in the two countries. America's prevailing attitude toward forests largely developed around the idea of "taming the wilderness" and opening it up to settlement. The American frontier was a dangerous place where Indians, outlaws and wild animals lurked until pretty recently. This environment has shaped our culture and probably explains much of our deep seated individualism.   It is difficult to imagine private landowners in America, for instance, giving up their right to keep people off their property.

On the other hand, the Germanic view of forests as places of refuge rather than ambush, goes back over 2000 years. The Battle of Teutoburg Forest in the year 9 AD was the decisive German victory over the Roman Empire which halted the advance of Rome. This battle is considered Rome's greatest defeat. It was a turning point in world history. The Romans never again invaded east of the Rhine River; preserving the German people as a distinct culture. The battle helped to romanticize forests and create a nationalistic pride and admiration for forests. What if the Battle of Teutoburg Forest had went the other way? What if it had been more like the last effort of the American Indians at Wounded Knee?

Only when timber became scarce did an American conservation movement get underway. America used what I call the "empire model" that has been followed globally by the Roman, Spanish, Portuguese, British and other empires. It's mode of operation is "Expand or die!" as the tiny Roman soldier on *The Night at the Museum* movie proclaims. The key to empire is controlling the transportation networks. Empires control the sea lanes and ports and/or overland trade routes. After WWII, America inherited Britain's naval network while Russia expanded their communist system across the vast Eurasian land mass. After the Cold War, it seems to me we are seeing a merger of the two systems: globalism. This new Empire model even further controls trade and extracts wealth from all over the globe at the expense of indigenous cultures. We see the same concept at work as big-box stores have put out "mom and pop" businesses here and abroad. The victims of empires have no choice but to join, become slaves, or die. A remnant may survive in the swamps, deserts, peatlands and other places not worth rooting them out…but they are too hard-pressed to fully maintain their way of life.

We will have to reject the "empire model" as our economic model if we are to save the primary forests that are still left around the world, preserve the wealth of indigenous knowledge that still hangs on, to keep our wonderful diversity of species alive on this planet, to even save ourselves. The only way this happens is through scarcity or ethics.

Gifford Pinchot believed that "The greatest good, for the greatest number, for the long run" was the ethic that should define forest management in America. Yet, different people have different ideas of what this means. I have stood among old growth white pines in the Menominee Indian Reservation and listened to their Chief Forester, Marshall Pecore, explain all the struggles his people endured to keep their forests from being plundered. Five times, they had to fight the United States Government in court. If they had lost, this magnificent remnant of what nature is capable of producing on these lands would have been liquidated. The few men who became wealthy from that would have been dead before the stumps rotted.

I stood marveling at the stand of massive white pines as an older gentleman began to take issue with Marshall's rationale. Then, I cringed as I listened to the career forester articulate his best argument that "the greatest good, for the greatest number, for the long run" would be for these trees to have been cut a long time ago. He said more money could have been made in the long run this way if the money had been invested along the way. His comment made my stomach churn.

I don't recall my German grandparents or any of the "old timers" who were raised the way they were as making decisions this way. Their goal was always, so far as treatment of the land was concerned, to leave things better off for the next generation than they had received it. This means materially and qualitatively; with full potential. Driving around my six-county District in Illinois, years later as a District Forester for the Illinois Department of Natural Resources, it was easy to tell which landowners were these "old-time Germans". Their homes were always modest but very tidy and well kept. Their vegetable gardens had flowers around them. Their yards had shade trees, fruit trees, and flowering bushes. Their forests were in good shape; not heavily cut-over and full of invasive species. They had high quality, healthy trees. Few, however, had been participating in any government programs to manage their forests. In fact, most had investigated this route and reacted about like the folks on the TV show "Green Acres" did to the Agriculture Extension guy…some "city slicker" that didn't know anything. What my Grandma liked to say about them being stubborn rang true: "You can always tell a Dutchman, but you can't tell them much" (Germans were commonly referred to as "Dutch" because their name for themselves is Deutch; pronounced "doytch").

"Seems to me like a bunch of educated idiots!" a few of these "Dutchmen" had told me, with a wink…fully knowing I was state employee myself. A few described to me how if they had been "falling for" the government programs all this time, they would have planted invasive species all over place. Sure enough, I found a seedling order form in my office from 1971 that was provided for participating landowners to fill out to obtain free seedlings for conservation purposes. All of them were non-native species. Most had since become invasive.

I was now trying to sign up people to get cost-share funding to eradicate them. If they had listened to the professional "experts", they said, their forests would have most likely been clear-cut by now too, and they wouldn't have any ready material for building barns or "money in the bank" in trees to sell "in a pinch" either. Standard forest management did advocate for even-aged management, which eventually results in a stand-replacement silvicultural clear-cut…which they wouldn't do.

Well, they had a point. The programs I worked with were good when compared to mismanagement or no management at all…but at best they could only succeed in achieving as much responsible management as could be paid for. Yet, these "old timers" just did what "made sense" without being paid…because it paid "in the long run" in the big-picture way of thinking. "High-grading"; which is cutting the best trees and leaving the rest, was often what other land-owners unwittingly allowed timber cutters to do to their forests. Far too many of them openly wanted to just "get all it's worth now", with no concern for the future. More often than not, the heirs of "tight" landowners who had held onto their mature trees were the ones to liquidate the timber once they "got their hands on it". Tax laws exacerbate the problem. The old-timers did what "made sense" and the others did what "made dollars".

Regardless of the financial decisions landowners made, I felt gung-ho to help them get their forests under active management and put back on the road to recovery if they were inclined to go that direction. But, I noticed that the forests I ran across that were already in good shape were mostly the ones these "old time" German-Americans owned and managed along the lines of what my grandpa had told me. I made inroads with them early on, and some of them signed up with our program. These landowners were all over 70 years old, but were still cutting firewood and farming. All I had to do was call was walk their land with them and point out what made sense to do, and they would get busy doing extensive timber stand improvement work themselves without receiving cost-share assistance.

During my six-month probationary period at my job, my boss was very con-cerned that my records showed far more acres treated than paid out in cost-share money. I explained that most of the landowners I worked with completed their projects themselves without receiving any assistance. He could hardly believe it! But, while he was impressed with this, he said that as far as the State is con-cerned, nothing happens unless it is tracked through money expenditure. So, he made me go back and get these people enrolled in cost-share assistance and I had to slow down to getting things done at the rate the money was available. Wow. Eye-opener number one.

So, my new strategy was to explain to landowners the 4% "harvest fee" on timber that came about with Illinois' "Forest Development Act". The act was created in 1983 to fund a program that would provide cost-share assistance to landowners to manage their forests. But, almost no one knew about it…even when they sold timber, it was largely kept hidden. Landowners who had had a harvest could apply for the funds and essentially get back their own money and then some…as not everyone who had a harvest participated. The money could be used to pay for timber stand improvement, tree planting, fencing cattle out of the woods and so on. Landowners could hire out the work or do it themselves and get paid the going rate.

Once I began using this strategy, landowners felt like they were getting ripped off by the government with this essentially hidden tax…and one after another of these landowners signed up to "get it back". Nearly all of them rapidly did the work themselves…my now, many were in their 80's. At the time, the forest management plans I was doing did not have to adhere to a very structured template and there was very little "fine print" in them or scary "legalese" stuff. I cranked out plans quickly and soon became one of the highest-performing District Foresters in the State in terms of numbers and acres of forest management plans written, timber sales administered, cost-share funds allocated, acres of trees planted, and so on.

The FDA program had been enacted in such a way that the timber buyer was held responsible for the payment of the 4% fee rather than the landowner, yet the money was deducted from the proceeds to the landowner of course. Timber buyers hated it and generally did not mention it to landowners because the landowners would get upset about this money coming out of their part of the proceeds and try to negotiate taking it instead from the timber buyer's part. If landowners inquired about how they could get the money back by participating in the cost-share program, they would find out that they had to have had a forest management plan and had a forester mark their timber…not just sell it to a timber buyer. Timber buyers didn't normally want that to happen, because it could mean a delay in them selling the timber, and probably competition from other buyers.

I had a few "hay day" years like this as a District Forester. There seemed to be no end to the level of getting things done that these landowners were capable of. It was very gratifying and encouraging. One 76 year old widow bought a chainsaw and implemented her seven acres of invasive species control while wearing a dress. She not only got it done, she stacked the cut brush into neat piles for wildlife cover.   But, none of these older landowners seemed to be able to get their children or grandchildren involved. They were too busy, lived too far away…or just didn't have any interest in working in the woods. The only landowners I am aware of who managed to get the younger generation to help them implement their forest management did so by giving them the ultimatum that if they wanted to hunt on the family land, they had to help manage it too.

Well, the "hay days" only lasted a few years. Several hits came in rapid succession to upset everything. First, the State handed down rules for us to follow that dictated a new "canned plan" template be used for all forest management plans and strict adherence had to be followed as far as having an inventory of the forest done (using a very antiquated and time-consuming computer program called IFIDAP: Illinois Forest Inventory Data Analysis Program). The really old plans, before computers, were only about four pages long and often had little to no inventory…just a written "eye balled" description of the forest areas. By the time I started, we were generally expected to have an inventory and we used Corel Word Perfect on the computer to do plans; yet, there was quite a bit of leeway in how we decided to do them. I would try to fit the plan to the forest owner as much as to the forest. I would get to know the forest owner and write the plan in such a way as I felt was most encouraging or applicable to that person. My first year doing it this way, every single landowner implemented their plan.

The new "canned plans", by comparison, were heavy on jargon, data, and "legalese". Mine ran about 27 pages long; mostly of bureaucratic "rigmarole". With the canned template, there was little leeway to customize them. It took forever to do the plans this way, and landowner didn't like them one bit. What had once been easy to understand…something we could sit on the tail gate and look at for a few minutes, agree to, and start implementing…became so convoluted it made you feel like you needed a lawyer to find out what you were getting involved with.

Next, maps had to be done with ArcMap GIS too. Previously, I just drew with a magic marker on an aerial photo that the county farm service agency would give out or a photocopy of the page in the soils map book. It took about five minutes to make a map for a management plan. We had to take classes to learn how to use ArcMap and then every year a new version came out that had to be learned; which slowed down the mapping from about five minutes to half a day or more just to make a map.

Then came Rod Blagojevich as Governor. His "administrative sweeps" swept away not only all of our cost-share funds, but in a brilliant accounting maneuver, he swept away the future funds too! We had a negative balance. The 4% tax still had to be paid, but no one would get any of the money. Well, Blagojevich eventually ended up in prison…but a lot of damage was done before that happened.

The next thing that happened was the "old timers" started passing away. Their heirs did not have the same interest and enthusiasm for managing their forests for the long term. Those who did have an interest in management were mostly oriented towards improved trophy hunting. Few had the time or interest in doing the work themselves, and plans took longer and longer to get implemented…if at all.

Finally, the Renewable Fuel Standard (RFS) happened. The Energy Policy Act of 2005 escalated the use of ethanol in gasoline. Ethyl alcohol, also called ethanol, is added to fuel not only to increase the level of "renewable" fuel used, but to oxygenate the fuel to increase the octane. It replaced methyl tertiary-butyl ether (MTBE) for oxygenation of fuel when MTBE leakage from underground tanks was shown to be contaminating groundwater and causing cancer. MTBE is derived from natural gas, and was itself a substitute for lead; which had been used for this purpose up until 1979…also because of deleterious health effects. Other additives exist, but are currently more expensive than ethanol.

The mandated increased use of ethanol resulted in the "ethanol craze". Burning corn for fuel caused market competition for finite corn supplies needed for human and animal food. Corn prices were rising fast, as ethanol production mandates pushed farmers to produce higher levels of ethanol…and thus corn. Lincoln Land Agri-Energy, LLC, was built near me in Palestine, Illinois…and many other ethanol distilleries went up to meet the ethanol demand. The environmental and political effects of the legislation are complicated and I won't get into all that here.

While Henry Ford's original engine was designed to run on ethanol, the corrosive effects of ethanol on vehicle engines, the fact that it dramatically shortened the shelf life of gasoline, and speculation as to "where are we going to plant more corn" was the talk of coffee shops throughout the Midwest. The effects on conservation were felt immediately. Some days, several farmers who were enrolled in conservation plans showed up at my office inquiring how they could "get out of their forest management plans". Many "paid their way out" at the Farm Service Center and began plowing up trees planted in the Conservation Reserve Program to plant these acres back to corn.

One particular farmer stands out. I tried to talk him into keeping his highly erodible and flood-prone fields in trees for the wildlife…and he looked me in the eye and said "To hell with the wildlife. I would have to be an idiot to leave my ground in trees if I can make even one dollar per acre more with growing corn."

Farmers all around me began to clear out fencerows and riparian areas, channelize and straighten waterways, and withdraw their land from conservation programs. It seemed like a horde of locusts had been released. The landowner bordering my property…which was enrolled in the Conservation Reserve Program and planted to trees…had just inherited the farm from his grandpa. He bulldozed out all the trees along our common property line; including those on my side, while I was on vacation.

LG drove up on his tractor one day while I was cutting up a cedar tree that had blown down in my yard. In his eighties, he was still strong and held himself like a much younger man…always at work around the farm doing something. He offered to move the log for me and load it onto a utility trailer. I planned to take it to an Amish woodworker to have it made into cedar chests for my children. After he got it loaded, I took off my chainsaw helmet with ear muffs and sat on the log. He got down off his tractor, shut it off, and walked up beside me. Standing there in his faded bib-overalls and seed company cap, he looked at the ground for a minute. He placed his thick, wrinkled hand on my shoulder and seemed to be looking for the right words to say. He looked around at the torn out, channelized creek and missing fencerow, then looked off across the road and stared into the distance.

"I guess I am the only farmer left in the county" he said; his deep voice rumbled, with a quiver in it. I could tell he was looking away to hide the tears welling up in his eyes. He still ran cattle, had chickens and a big garden, and farmed. He went on to explain how he worried how his son was already borrowing large sums of money to buy new equipment and a semi-truck to get geared up for taking over his farm. "This old tractor and a grain wagon has worked fine for me, for years…" He was worried this would lead to him getting so far in debt chasing money that he would first ruin the farm and then lose it too.

I was burnt out and ready to get out of there. All this gung-ho effort and apparent success in improving forest management and land conservation was rapidly going down the drain. The view from my home was ruined now and the creek channelized as it ran into my small property. Pesticide drift ruined my garden, killed all the grapes I had planted, and tattered the leaves every spring on my white oak trees. I put in for a job transfer to extreme Southern Illinois… where the landscape was much more heavily forested. My supervisor approved it, but *Blago* never signed the transfer; a routine procedure that, before him, would have taken two weeks. Two years of limbo passed with not knowing where my job was going to be or if it would be eliminated.

Clearly, for any lasting results to happen, a "land ethic" had to come about. The new generation of landowners who came after the "old timers" would work like the dickens to make one more dollar per acre, but had little motivation or even a thought to do much of anything for the greater good of the community or the environment. The civic clubs, like Ruritan and Kiwanis had been mostly comprised of the older generation too…and began to dwindle. How did the "old timers" develop such a strong "land ethic", a strong work ethic to put it into action, and a sense of community responsibility that persisted throughout their lives?

The older generation that grew up during the Great Depression were shaped by their culture and community to do something worthwhile for their community, to give something back. This was noticeable among all ethnic backgrounds. However, nearly all of the landowners who took a hands-on interest in managing their forests were of German descent and were educated in a small, rural Lutheran or Catholic church-based schools. Their way of doing things on the land, however, was rarely passed down to their 'Americanized' offspring. I wondered…was such a "land ethic" still intact in Germany…even though they had been through the era of Nazism and the devastation of losing two World Wars? Did they still love to work in the woods like these old timers did, even in their old age…so motived to "get stuff done", even if they would not live to benefit from it?

One day, I received a call in my office from a student who was an intern in Freiburg, Germany. This is before I got to go there myself. He was from Pennsylvania, and had come across my name somewhere and was inquiring about employment after his return to the USA. I talked to him at length about his experience there. He worked for the City of Freiburg in their community forest. He said it was incredible…it seemed that everyone enjoyed hiking and walking in the forest above all other activities, and even enjoyed working in the forests. He told me there were forest trails everywhere and people could roam the forests all they wanted…there was no such thing as trespassing…the forests were open to everyone and yet no one trashed them, littered, or stole artifacts.

"Steal artifacts?" I asked him. "Oh yes", he said "There are these shrines and all sorts of pieces of art and wood carvings along the community forest trails… I even saw an ornate cross that is hundreds of years old…and yet no one messes with anything." He related how families would go out on the forest on weekends and cut firewood on the community forests that had been cut down by the foresters in thinning operations and carry it long distances by hand to get it to their vehicle…how there were community apple orchards and families would go out and pick apples and make apple cider and wine in such large quantities as to last all year, and described the beautiful community gardens they worked in… full of flowers and vegetables…so many people working doing these things and so happy to do it. He was young and gung-ho about bringing this kind of activity back with him to the USA.

My boys were briefly in the Boy Scouts during this time, and I met an active Scouting dad whose wife was from Germany. He said his brother-in-law was coming in to visit and wanted to get out and see forests here. Knowing I was a forester, he asked if I would mind taking him with me to work for a day or two. I took him along the next day to inspect some timber sales I had administered. He was a factory worker with no formal training as a forester, but knew more about forestry than a lot of foresters I have known. He wasn't a fluent English speaker, but we could understand each other. He appreciated that I knew a little bit of German. He showed me photos of the community forest he and his factory co-workers cut firewood in and the wood pile they worked up while on break at the factory. This was done as a joint effort, with them all working together and sharing everything. The trees were cut as part of the management of the forest. Their firewood stack was the size of a train car and so uniform and neatly stacked that it looked like a solid block of wood.

We arrived at the forest I had come to inspect. It was in an oil field area and we parked at the end of a lease road. Getting out of my truck, we immediately saw empty beer cans scattered all over in the weeds. This was so routine for me, that I just walked past the cans. But, he was so upset by the trash, he started picking it up and seemed to be swearing in German. We cleaned up the cans and he asked me why anyone would do such a thing. I explained that it was probably teenagers who had come out here to be hidden to drink beer…because here the drinking age was 21. He hit himself on the head with the palm of his hand, made an exaggerated expression, and said a German word I didn't know the meaning of. Then said, "In Germany, no one would care if they drank beer…only that they throw out trash."

We went into the woods. The harvest was typical for what I had come to expect from years on both sides working in the timber industry. There were some ruts, skinned up trees, top logs that could have been skid out but were left, and so on. It wasn't pretty, but it wasn't bad; an average timber harvest.   To him, it was horrible. He looked at me as if I should be jumping up and down swearing and asked "What will be done with all this wood that is left? These size pieces would be used by the local woodworkers." He even picked up a piece about two feet long and commented that it would make good clock wood and be sold. I told him probably nothing…there weren't really any local manufacturers around that made anything out of raw wood like that or anyone to sell it to except for firewood. He could hardly believe that either…no local markets or manufacturers?

There it all was…the missing ingredients we were lacking here: Thoreau's *Economy, Leopold's Land Ethic and Community.*

I finished my work and we headed back to the truck. When we had arrived, the beer cans had distracted us from noticing the oil well pump jack nearby. Like almost all pump jacks I had been around, and I had been around been leaking oil somewhat from time to time, and the soil around it was soaked with oil. The lease roads had oil spread on them to keep the dust down, and there were some "salt kills" around from salt water leaks. This was all normal to me. He couldn't believe such messiness was tolerated and wondered about the water quality in the nearby creek. I told him about the land my Grandpa had owned, and how it had the same issues. The creek had been severely impacted from salt spills over the years and hardly had any fish or mussels in it anymore…but when he was a boy, you could fish and swim in it. Like most landowners, he didn't own the mineral rights and didn't get any money from the oil coming out of the ground, but had to put up with the damage caused getting it out. Then, someone stole the timber too.

He asked me where I went to eat during the work day, as he hadn't seen any restaurants. We had passed through several small towns and rural crossroads with their vacant stores, vacant restaurants, and vacant schools. Only churches remained…most with only a handful of elderly members still attending. I told him I usually took my lunch to work and sometimes stopped to eat it at one of these many country churches that are everywhere. I told him that they nearly always had a bathroom and most had a little kitchen area. I would clean up, warm up my lunch and refill my canteen. He nodded in approval. That's the only thing I showed him that day that he didn't have a negative reaction to. When I told him there was no *Bierstube* anywhere around…much less a *Biergarten,* he gave me a look and shook his head back and forth. He couldn't believe either Prohibition or drunk driving could exist…let alone in the same place.

My favorite church to stop in for lunch was in Jasper County. It had apparently been abruptly abandoned some years before as there was no sign that anyone had been inside for years. If there ever had been a parking lot, it must have been plowed up and incorporated into the farm field beside it. Dusty hymn books and bibles were still sitting in their slots behind the pews. The building had never been updated to have plumbing or electricity, and two outhouses stood behind the church building. I brushed the dust off the men's seat and put a roll of toilet paper in it for myself. The churchyard had grown past the weed stage and was now regenerating into saplings. My beloved Embarras River was nearby, and when it was out of bank, the route to the church was cut off. This was the last church I stopped at while working in Illinois…to say goodbye…not to the building, but to something else that seemed to still be there after everyone was gone.

That was twelve years ago. I hope the church has not been since tore down to make a few more rows of soybeans. I had fancied the idea of trying to buy it, as my Grandpa had his old one-room schoolhouse, and fixing it up. It looked a lot like it. It seems that the old country churches are disappearing too. If the fellow from Germany was to visit now, I would be hard pressed to even find an unlocked church to stop in…now that meth and heroine have invaded our rural areas.

Wendell Berry understands and said it well:

"People use drugs, legal and illegal, because their lives are intolerably painful or dull. They hate their work and find no rest in their leisure. They are estranged from their families and their neighbors. It should tell us something that in healthy societies drug use is celebrative, convivial, and occasional, whereas among us it is lonely, shameful, and addictive. We need drugs, apparently, because we have lost each other."

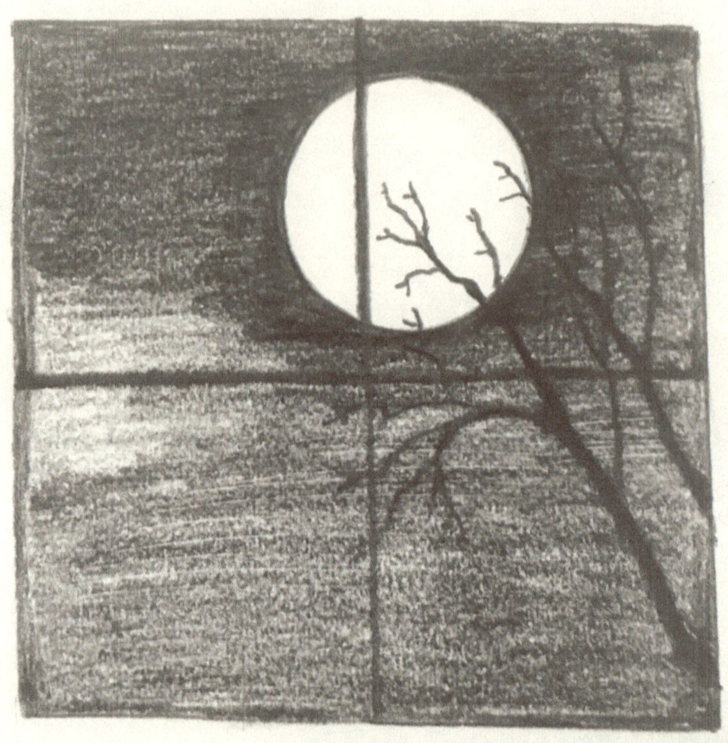

Chapter 7:
Rest

Goethe's "Wandrers Nachtlied" in German (1780), inscribed on the wall of a
wooden hut on the Kickelhahn mountain, near Ilmenau.
Über allen Gipfeln
Ist Ruh,
In allen Wipfeln
Spürest du
Kaum einen Hauch;
Die Vögelein schweigen im Walde.
Warte nur, balde
Ruhest du auch.

Written in 1780, here is a translation by Henry Wadsworth Longfellow (1845):

Wanderer's Nightsong
O'er all the hilltops
Is quiet now,
In all the treetops
Hearest thou
Hardly a breath;
The birds are asleep in the trees:
Wait, soon like these
Thou too shalt rest.

The The short poem, "Wanderer's Nightsong II", has been described as the most perfect lyric in the German language. Scribbled on the wall of an old wooden gamekeeper's cabin on top of Kickelhahn Mountain in 1780. What makes the poem such an influential piece is that in one small piece, it unites landscape, creation, and all beings together in evening silence…and places man as restless but expecting sleep…in death and eternal peace. Goethe "wanders the cosmos" in this little poem, and it was perhaps the first poem to unite these concepts this way. The words are more fully understood in the original language. The word for "rest" used in the final line is expressed as a verb (ruhen) and not a noun (Ruhe). This verb form suggests that, even in "rest" there is movement. The last word of the poem is "auch" (too, also); which further emphases the kinship between humans and the other creatures in Nature. All of Creation, and even the Creator rest:

"Thus the heavens and the earth were finished, and all the host of them. And on the seventh day God ended His work which He had made; and He rested on the seventh day from all His work which He had made." – Genesis 2:1-2

The seven day work week has its origins in ancient Judaism…continued into Christianity and adopted officially by the Romans (who formerly had an eight day week) by Constantine in AD 321. The influence of both Rome and Christianity made it spread pretty much throughout the world. The pre-Christian, pagan influenced the naming of most of the days of the week…with Norse versions of pagan gods eventually winning out for the most names. Monday, "moon day" after the moon; Tuesday after 'Tiu', a lesser known god of war in the English/German pantheon; Wednesday after 'Woden', another name for

'Odin', the chief god of Anglo-Saxon mythology. Thursday is named after 'Thor', the Norse god of thunder and lightning (associated also with Jupiter in Roman mythology and Zeus in Greek mythology). Friday is derived from 'Freya', the Norse goddess of love, marriage and fertility (Venus in the Roman, and Aphrodite in Greek). Saturday, in English, is from "Saturn's Day"…Saturn being the Roman god…while in Spanish and French it is called "sabado" and "samedi" and refers to the Jewish Sabbath.

Sunday, is "Sun's day", or "day of the sun"; translated in both Latin "*dies solis*" and Greek "*hemera helio*" as "day of the sun". In Spanish and French, however, Sunday is "*domingo*" and "*dimanche*"; again referring to "Sabbath day" but here more literally translated as "Lord's day"…pointing to the Christian God. So, in the naming of the days, the pagan, Jewish, and Christian perspectives were all preserved.

Regardless of the number of days of the week or the names of the days…rest is needed. I have heard accounts of the wagon trains which traversed from St. Louis, Missouri across the frontier all the way to California…and how the wagon trains which stopped to rest on Sunday made it there quicker and in better condition that the ones that attempted to push through every day without rest. Thankfully, most of us reading this book are able to have at least one day of rest per week…and hopefully enjoy a work week limited to 40 hours. I saw a bumper sticker the other day that said: The Union Movement. From the folks who brought you 'the weekend'".

My Grandpa Torbeck was a Teamster…and represented his small group at Feller Oilfield Company, where he drove an oil rig truck for about 50 years. One time I was riding with him in his oil rig, when I was about 13, and he pulled up to a creek bridge and ran the hose down into the creek to suck in water to fill the tank. Standing around waiting for the tank to fill up, he was smoking his cigarette and complaining about how everything is "Republicans vs Democrats" and nobody could figure out how he could be both a Teamster and a Republican. "The Republicans would have things run without any consideration for the worker, and the Democrats would have things run without any consideration for the employer…ya gotta have a balance…jobs are going to just go overseas if you have too much of one thing, and nobody can make a decent living if you have too much of the other." He repeatedly was chosen to represent the Teamsters in his little area for being level-headed and reasonable. It's all about balance.

Rest. This last chapter, titled "rest" is more about renewal than relaxation. Without rest, we do not have time to think…to ponder…to meditate. We do not have time to heal physically, mentally, and emotionally. Before our time of final rest…death…it is imperative that we have this time or else we have not fully lived. Some of us call this time to rest "free time"…and that is how I like to think of it. We all need time to be free to think, contemplate, reflect, and get ideas for our next project. Before making a big decision, often people will say "Let me sleep on it"…and for good reason. Sleeping, we dream. Our brain is not able to tell the difference between the dream and reality. Often, I get my most important "ideas" from dreams. Ancient people have tried to interpret them for time immemorial.

William Wordsworth, the great English Romantic poet had a startling concept: that the life we live is a dream and that we 'awake' to 'reality' when we die…somewhat reminiscent of the "Allegory of the Cave" by Socrates.

'Our birth is but a sleep and forgetting:
The Soul, that rises with us, our life's star,
Hath had elsewhere its setting,
And cometh from afar.'

Similar concepts have been described in Buddhist writings. Regardless of the significance of dreams, we need rest, relaxation, and sleep to be productive in our waking lives. Working with horses, a person has to slow down to accommodate the horses in a way that does not have to happen when working with machines. Thus, the machines that were supposed to give us more rest and relaxation have, in many cases, driven us to a degree of business not experienced before. This makes it all the more important for us to be able to make time to rest and relax. We would do well to create a relaxing atmosphere or habitat to promote a sense of peace, security, and harmony so we *can* relax. Livable landscapes promote relaxation and reduce not only stress, but violence and crime as well. We need to have beauty and green space around us to feel healthy and act helpful.

Pearl Fryar created such a place on his three acres in Bishopville, SC. Fryar's topiary garden has become a major tourist attraction in this otherwise non-descript little town. Pearl, an African American, is a perfect example of how creativity combined with "a little know how and a lot of elbow grease" can create something of beauty and utility that can transform a community. He often shares his story with area youth and encourages them to work hard to achieve their dreams. He graciously welcomes people to see his property, learn his methods, and use his fame to draw visitors to the town. Pearl's magnificent creations have dumfounded horticulture experts and inspired countless youth to follow their dreams. Imagine if every community had a Pearl Fryar.

———————

"The world is but a canvas to our imagination." – Henry David Thoreau

Frederick Law Olmsted, who famously designed New York's Central Park and the U.S. Capitol grounds, is known as the founder of American landscape architecture and the nation's foremost park maker. Olmsted was born in Hartford, Connecticut, in 1822. He worked a variety of odd jobs ranging from clerk to farmer, and was even a sailor in the China trade for one year. Olmsted was also quite a wanderer and writer. He traveled throughout the American South in 1850's and gained fame for books he wrote describing the slaveholding culture of the South.

Next, Olmsted traveled to England. Wandering the countryside, he became impressed with the English landscaping. Describing what he saw, he wrote the book "Walks and Talks of an American Farmer in England" in 1952. Now, at age 30, Olmsted had found his passion: landscape architecture.

In 1857, without having any college education, Olmstead was hired as the superintendent of New York's Central Park during its design phase. Together with fellow designer Calvert Vaux, he won a competition to design the park, and became its chief architect in 1858. Working together with Vaux, and then later on his own, Olmsted designed many of the nation's premier urban parks. He also continued to write, and co-founded "The Nation" magazine and the journal "Garden and Forest".

Restless and ambitious, Olmsted got involved with many efforts to preserve natural beauty. He served as the first head of a commission charged with preserving the Yosemite Valley and together with Vaux, helped to establish the Niagara Reservation to preserve the area around Niagara Falls. He designed many of Louisville, Kentucky's old, grand parks as well. Olmsted also managed to find time to serve as the administrative head of the US Sanitary Commission; which went on to become the American Red Cross and he served as manager of the Mariposa gold mining estate in California as well.

By 1892, Olmstead was hired to design the famous Biltmore Estate grounds in Asheville, North Carolina. The next year, he served as director of the World's Fair in Chicago. He retired from his work in 1895 and, perhaps from so many years of such intense work and drive, he suffered a mental breakdown. Sadly, he spent the last few years of his life resting in McLean Asylum in Waverly, Massachusetts and died in 1903. (Information from www.fredericklawolmsted.com)

Olmsted's legacy lives on today. His innovative, urban landscape design and nature preservation are becoming more fully appreciated today as folks rediscover the greater quality of life to be enjoyed by carefully planning our urban environment with, rather than against, nature. Olmsted's lifetime of accomplishment is a great example of what is possible when self-education and determination is applied to our passions.

I find it ironic, however, that the Vanderbilts would go to such great expense to beautify the grounds of their immense mansion…but did not seem to be concerned about the aesthetics of their vast forest holdings. The forest was expected to generate money, while the home and grounds is where great sums of money were spent. Yes, they employed professional foresters to manage these lands…but I wonder what results could have been achieved if the forests had been managed with an eye for aesthetics and other such matters of the arts as well.

St. Louis' Forest Park is where I was first exposed to landscape architecture. When my transfer within the Illinois Department of Natural Resources was foiled by crooked politics; leaving me in professional limbo for two years, it was my early exposure to Forest Park which inspired me to consider venturing into urban forestry. Two years as an urban forester in Clarksville, TN afforded me an opportunity to shape the urban landscape there. An increasing majority of the world's population now live in urban areas, and this makes urban forestry an issue of importance to all foresters. The public's impressions derived in an urban setting will thus increasingly translate into how society views forests as a whole and allows foresters to manage them. If we want to see forests managed well "in the boonies", I think it is therefore imperative that beautiful, purposefully built landscapes be developed in urban areas. So, I would like to introduce one of my favorite landscape architects: Jens Jensen. Again, my upbringing in the Midwest is responsible for this exposure.

Many of Chicago's famous parks, as well as the Lincoln Memorial Gardens in Springfield, Illinois, were designed by Jensen. Like the more famous Olmstead, Jensen was also involved with nature preservation efforts at an early time in our nation's history. He organized conservation efforts across Illinois and Indiana, which led to the creation of the Cook County Forest Preserve District, the Illinois state park system, and the Indiana Dunes State Park and National Lakeshore.

Jensen was born in 1860 and grew up on a farm in Denmark. He immigrated to the United States at age 24 and quickly developed a great admiration for the natural landscape of America. He worked in Florida and Iowa; then landed a job as a laborer at a city park in Chicago. Rising through the ranks, he was soon designing Chicago's parks and in his spare time he travelled extensively throughout the Midwest photographing the landscape. Jensen is perhaps best known for his "Prairie Style" landscape designs which incorporated the use of native species and materials. Long before "going green" was a part of our vocabulary, Jensen was extolling the renewing and civilizing powers of nature and "living green".

Henry Ford was a big fan of Jensen, and hired him to design his estate in Michigan. It may seem ironic that Mr. Ford; who brought the automobile to the masses, was interested in Jensen's "living green" philosophy. However, Ford actually built biodegradable cars made using hemp and the cars were made to run on vegetable alcohol. As early as 1924, he was advocating the use of bio-fuels, stating: "There is enough alcohol in one year's yield of an acre of potatoes to drive the machinery necessary to cultivate the fields for a hundred years." (greenlivingideas.com)

In 1935, at age 75, Jensen retired from landscape architecture and founded The Clearing, a "back to nature" folk school, in Door County, Wisconsin. Jensen went on to help establish a nature sanctuary and several parks in Door County before he passed away in 1951, at the age of 91. Jensen believed that an understanding of one's regional ecology and culture helps to develop "clear thinking", while maintaining our contact with the soil provides a basis for our values. The Clearing still exists today, and continues to provide hands-on classes in natural sciences, fine arts, skilled crafts and the humanities. (theclearing.org)

Industrial hemp was once a major crop in Kentucky. As with the demise of Prohibition, hemp is again becoming legal and new hemp-manufacturing companies are starting up in Kentucky. I wonder, if Henry Ford had been able to realize his dream of manufacturing his car panels from hemp and using alcohol to power them, where we would be today? He hoped for replacement parts to be made easily by locals from biodegradable hemp and the fuel to be produced locally as well…all by farmers. There would be little need for auto salvage yards, gas stations, or oil wells. I think about this every day when I drive the short distance from the Berea College Forest to my home just outside Berea. There are almost as many "junk yards"…auto salvage yards…as pharmacies here. I drive past two of them every day as I travel to my home from work. There are also three storage building centers within a mile of my home. I remember the first time I saw one of those…I never dreamed they would spring up all over like they have. There is so much stuff, so much junk…

Aldo Leopold realized that his concept of a land ethic would not become a reality until a consumption ethic had developed. For instance, in 1928 he wrote: "A public which lives in wooden houses should be careful about throwing stones at lumbermen, even wasteful ones, until it has learned how its arbitrary demands as to kinds and qualities of lumber, help cause the waste which it decries." In other words, when we demand the preservation of natural resources; yet at the same time demand more of those resources for our consumption, we are counter-productive.

Take, for instance, the comparison of American consumption in the 1950's to today. Often the 1950's in America is considered the "good old days", when the "American Dream" was possible for most everyone. These were the days of "Leave it to Beaver". Consider, however, that in the 1950's, the average home was 1,100 square feet, while the average household had four children. Now, the average home is well over 2,000 square feet, while the average household has fewer than two children. Obviously, our standard of living; and hence our consumption per capita, has increased dramatically. We consume more than twice as much, but are we twice as happy?

I grew up in the 1970's. I remember when most families still made it on one income and often owned just one vehicle. Only "rich people" had a riding lawn mower, bass boat, or ATV. Most children shared a room with a brother or sister, and it was a big deal to go out to eat once a week or so… usually to Pizza Hut. If a teenager wanted their own car, it meant that they would bale hay, de-tassel corn, cut tobacco or mow lawns until they could buy an old clunker by themselves.

Each generation strives to give their children more than they had. This is a good thing. However, perhaps the current economic situation is revealing to us that there is a limit to this. We can now see that much of the increase in the standard of living has been built upon debt. It has been an illusion. Richard Feynman, the famed American physicist, once described how there were a hundred billion stars in the galaxy and then remarked: "That used to be a huge number. But it's only a hundred billion. It's less than the national deficit! We used to call them astronomical numbers. Now we should call them economical numbers."

Today's generation hopes to give their children more as well. But, rather than more material things, I think many of us hope to give our children more family time, cleaner air and water, healthier food, and a more intact environment. Cities are catching on and beginning to consider their open space and urban forest as part of their essential infrastructure. They are providing walking trails, farmer's markets, and recycling opportunities. Perhaps Leopold's "consumer ethic" is beginning to blossom. Certainly, we want our economy to rebound…but not from increasing consumerism and debt.

The American dollar is often referred to as a "green back". How we spend our dollars can influence the "greening back" of the world around us… perhaps even more so than how we vote. Even Thoreau recognized this when he wrote

Walden in the 1840's; which begins with the "Economy" chapter. I read Walden on the way to Connecticut when I was fourteen. My mom and I took the trip together and our destination was New Mystic. A friend of hers was taking care of an elderly lady whose home was right on the rocky coast. There was a pencil drawing of an immense white oak hanging on her wall. Naturally, I asked her about it. The old lady, nearly one hundred years old, got teary-eyed when she recalled how stately the tree was, and how it finally died. She had played under it as a child, and loved that tree.

My mom let me pick the route to Connecticut. I navigated us to Niagara Falls, across into Canada a bit, through the Adirondack Mountains, and the southwest corner of Vermont. I kept some Canadian money for a souvenir...my change from buying a fake-fur stuffed baby seal. At the time, the beautiful white baby seals were being clubbed to death and this was a big news sensation that was getting attention. People wanted the practice banned, and yet some of the Canadians told me this and fur trapping was how the Natives made a living...without it, they would end up selling oil leases and we'd have worse environmental problems. Contemplating all this, I was on the Economy chapter in Walden, and using a Canadian paper dollar for a bookmark. It had a young Queen Elizabeth on it. The other side had a scene showing logs being floated down the river at Ottawa. The coins had a maple leaf, a beaver, a sailboat, a loon. The dollar coin, with a loon on it, is nicknamed a "looney".

"Seems pretty looney", I thought to myself as I compared it to American money. At fourteen, I began to earn it myself and I paid more attention to it now that taxes had been taken out of my earnings. The bills said "In God We Trust" on them, but had pictures of presidents and government buildings on them too. The small print indicates it as legal tinder, but each bill tells which Federal Reserve Bank...whatever that is...printed it. It's black on one side and green on the other, it's a rather schizophrenic piece of paper.

I liked the maple leaf and animals on the Canadian money. Considering the picture of Queen Elizabeth made me wonder what the USA would be like if it, like Canada, had not fought for independence against England. Canada didn't seem so bad off...it just had this young queen on its money instead of our presidents. I guess the Revolution was mostly over taxes, so it seems the money is about the only difference I could tell between the Canadian and American side of Niagara Falls besides the view...and more A&W Root beer restaurants than McDonalds.

The experience of exchanging my American money for Canadian money evoked a memory that affected my curiosity towards the "money system". I was seven or eight, playing with my Tonka dump truck on the concrete floor of the basement while Grandma was washing and wringing out "Uncle Charlie's" clothes. Carl, or "Charlie" as we called him, was the quintessential "old timer"... one of the several "old bachelors" as Grandma called them, who lived out on the

German prairie pretty much unchanged since before WWI. I loved to go along with Grandma when she would run stuff to them from the store. She would drive past the old Schaefer store and the old St. Paul store…now abandoned buildings…and into St. Peter, where there was still a store (that one is gone now too) and pick them up some staple items and beer.

Charlie had been engaged in his youth, but the young lady died before they could be married. It seemed his heart did not recover. He never married and never changed with the times. He lived down by Flat Creek in Grandma's old home place…making fishing seine nets, whittling, gardening, and living off the land the same way folks had done when he was growing up…all the way up into the 1970's. Grandma was washing his clothes and talking about taking them to the nursing home to give away. She kept his WWI uniform. The green wool leggings and hat are sitting in my home.

"Oh fiddlesticks! What in the world!?" Grandma exclaimed as Charlie's shirt got stuck in the wringer. Grabbing a pocket knife off a nearby shelf, she cut the seam and out fell several coins. "By golly, that Charlie, he never did trust banks! Shoot, I wonder if there's any more in these clothes!" An aura of excitement filled the air.

I quickly left my Tonka truck and picked up the coins; handing them to her as she continued to wring out old shirts and bib overalls and extract coins and carefully folded bills. As she did this, she explained how when the banks "folded up", folks lost their money, and he never trusted them or used them after that. The coins were old…some had different images on them than what I had seen. Bicentennial quarters had just came out, and been pointed out to me, so I was intrigued with coins. She pointed out one had an Indian on it and another had a Mercury head on it.

It wasn't fake pirate money either. Long John Silver's was my favorite place for her to take me and my cousins to eat, as we'd all get paper pirate hats and fake gold bullion. We'd then go to the Vandalia park where there was a huge uprooted oak tree to play on and pretend it was a pirate ship. Maybe it was real pirate money? "Are there any gold coins!?" I excitedly asked as she slipped another coin out of a shirt cuff.

"No, Roosevelt made everyone turn those all in. That's another reason Charlie didn't trust banks…or the government." "Who's that?" I asked. She explained that he was the president that got us out of the depression and through WWII. Her excited and happy demeanor changed to a solemn quiet and she looked down into her big apron pocket at the wet money. "Just a handful of coins and a few crumpled bills…not worth much…about twenty dollars", she said, her voice trailing off as her posture slumped and she headed upstairs…patting me on the back as she walked past.

I couldn't understand Charlie going to all that effort to sew coins into his clothes if the whole lot of them was only worth about $20. I learned, years later, that the same year (1933) the Civilian Conservation Corps that Grandpa had worked in was also the year FDR signed Executive Order 6102; which required gold to be surrendered. People were paid $20.67 per ounce for their gold in Federal Reserve Notes ("money"). Immediately following the surrender period, the Gold Reserve Act of 1934 raised the price of gold to $35.00 per ounce. This effected an immediate government profit of $14.33 per ounce. No wonder the government had the money to pay people to work in the CCC's. I realized then that sewing money into our clothes, or burying it in a jar as some others would, didn't keep it from being stolen. When Charlie sewed those coins in his clothes, they had far more buying power than when Grandma took them out.

Fortunately, Charlie didn't need much money. He rarely bought anything. He could make or grow just about everything he needed; just like everyone did out there when he was a young man, before the World Wars. But, back then, there were general stores, schools, blacksmith shops, sawmills, lumber yards, grain mills, apple orchards, and dairies out there…a community. Cattle, hogs and chickens roamed on nearly every farm, outside. Farmers saved seed to plant the next year. Gardens were planted with what are now considered heirloom varieties…many of which are in danger of being lost. He could putter along and get by, alone, as the local economy disappeared…until he passed away too. Everyone else had to "get with the program" to get by. Farmers had to "get big, or get out". The ones who "got out" went to work in the oil field; the main employer in the area during this transition. Carter Camp and other oilfield camps…much like the company coal towns in Appalachia, sprung up to house workers…moving them in from the hills of Oklahoma and Kentucky. Others would move to Granite City, Decatur, or Chicago; like many of our relatives did, to work in the steel mill or a factory.

Perhaps the most prophetic communicator on the topics concerning our money economy, the land ethic and the consumer ethic today is Kentucky farmer and writer Wendell Berry. He does an excellent job providing an alternative way of thinking about our economy that I think Thoreau would be happy with. Local economy is one of his primary topics. The more local the economy, the more closely tied are the people to each other, the land, and its resources… taking better care of both. According to Berry, "The care of the Earth is our most ancient and most worthy and after all our most pleasing responsibility." Wouldn't it be great if we can build an economy around the care of the Earth's ecological and human community rather than the destruction of it?

The key to rest is finding the time to do it…and one component of that, is doing things as efficiently and correctly the first time so you don't have to do it over. Doing things at the right speed, and thinking it out, comes to mind. This is why experienced carpenters say: "Measure twice, cut once". Everywhere, if we take a little time to "measure twice" before doing something, we improve the operation. This concept can be incorporated into large operations.

I drove a Toyota Tacoma for my work truck at Berea College for several years and many people in our area work at the Toyota plant building Japanese cars in America."

Most analysts attribute Toyota's success to its strategy of kaizen, which is Japanese for 'continuous improvement' and which just as easily could be called corporate deep practice. Kaizen is the process of finding and improving small problems. Each employee, from the janitor up, has authority to halt the production line if they spot a problem. The vast majority of improvements come from employees, and the vast majority of those changes are small: a one-foot shift in location of a parts bin, for instance. But they add up. It's estimated that each year Toyota implements around a thousand tiny fixes in each of its assembly lines, about a million tiny fixes overall. The sign over the door of Toyota's Georgetown, Kentucky, factory puts it in perfect deep-practice language: 'When something goes wrong, ask WHY five times.'" (Coyle 2009)

This practice of continual improvement applies, I think, to the forestry. How so? Well, in my experience as a forester, if a forester notices a reason to officially change something "on the ground" in the management of the forest under his/her responsibility, it is quite a cumbersome process to do that. Typically, management plans are done based on data collected combined with established silvicultural methods approved by the forestry academic establishment over many years of research and publications. It results in a "color by numbers" painting rather than a "freehand" painting…to use artistic terms.

When we "measure twice" as a carpenter, we don't measure all the boards before we start building…we measure them as we are building. When I was managing my own forestland I used to own…I would begin implementing the plan before I finished writing it. This might sound backwards, but I reasoned that all the tiny little decisions that go into whether or not to harvest a tree, thin a tree, plant a tree and so on…can be made so much more accurately on the ground in that exact spot where it is being observed, rather than in an office a week or more later while looking at numbers on a computer screen. If multiple foresters collected the data, no one person saw the whole forest up close. This is exacerbated further if the forester collecting the data isn't the same person writing the plan or implementing the plan. You end up with "factory forestry"…instead of silvi (forest)-culture.

Plantation forestry is fine to manage in a "factory forestry" method. Trees are planted as a crop and everything is uniform and literally "done by the numbers". The process is routine, as the rows of even-aged trees are not a complex system. But, ecological forestry…forestry practiced on native forests… should not have to fit a model that was developed for simple even-aged stands and is determined by "cookie-cutter" means. Increasingly, as we learn more about forests, it is resilience and structural complexity that is wanted; not homogenous stands. While using one forester from start to finish may not be practical all the time, large operations could be done in such a way as to allow a kaizen process so that the tiny decisions for improvement can be incorporated into the final plan and implemented as such.

How? My suggestion is to focus on the narrative rather than the data. Instead of focusing on the minutia; focus on the big picture. Tell a story. The original forest management plan for the Berea College Forest contains virtually no data…no inventory.   It reads like it was written from notes the forester took while riding through the forest on horseback. Many of the early plans that are considered pioneering efforts in forestry are just descriptions and narrative that guided experienced foresters on the ground to implement the general goals of the head forester. A well-written narrative can convey more than numbers…it can convey the attitude, the passion, the vision…that the person or people in charge of this forestland wished to preserve or create. One hillside may convey a totally different forest stand than the next; but the vision for the forest as a whole is what the foresters focused on. We should to.

Many forestry terms could be changed in our narrative plan to improve the vision for the forest. For starters, forest management instructions are called "prescriptions". The word "prescription", to me, implies sickness. Imagine if we used the word "recipe" instead. Next, the practice that is prescribed is called a "treatment"! Ugh. Often, the case is that the forest is determined to "need disturbance". Ugh again. So, we are going to disturb our sick patient because we have determined he needs a treatment…not a very upbeat story to tell. And, we have to tell it using a lot of numbers and terminology that the average landowner doesn't normally use.

Let's imagine the forester not as a doctor who prescribes a medication to a sick patient; but as a chef who uses the ingredients provided by the forest to produce a fantastic assortment of healthy recipes for a community dinner. The forest is a group of good friends. We will assure our forest friends, as we serve them; that they will receive our best efforts to insure that they and subsequent generations of their family thrive and prosper. Our community is treated like family and we are all members of this healthy, functioning ecosystem that is the community all of our families live in. Better terminology could help the forestry profession communicate a more positive image to the public. More importantly, it could help stimulate a more creative process for us as managers too.

This takes imagination and vision. What is your vision for this forest? When we think in these terms when we write a forest management plan, it focuses us in two ways: One, we use our "mind's eye" to creatively visualize what we want the end result of our management to look like and thus become in reality. Secondly, it takes our focus off of what we are tempted to perceive as

"wrong" with the forest and instead focuses our attention on the myriad possibilities the forest affords. We focus more on what trees we are leaving, what
ecosystem functions we are leaving undisturbed, and what ecosystem functions
we are promoting…rather than what we are removing.

––––––––––

Rest and learning go hand in hand, and the rest of an active person is not
wasted time. Our rest time may even be some of our most productive time.
Americans might marvel at the large amount of time off that Europeans get.
When I was planning my trip to Germany, at first I thought to go in August.
But, when I inquired about arrangements, the forester told me on the phone:
"No one works in August…I will be in Corsica hiking and the other foresters
will also be on vacation." Indeed, many work less than 40 hours a week, get
generous maternity and paternity leave, and very lengthy vacations…yet still
receive a good salary. No wonder every time I go on vacation somewhere, I see
disproportionate numbers of Europeans vacationing. The learning that occurs
and the relaxation that the spirit, mind and body enjoys, is beneficial to their
employers and to the society at large. Becoming invigorated through "free time"
is recognized there and productivity is enhanced rather than degraded.

This is not a new concept. The early Greek word for leisure was "*scole*". This
word forms the root of the English words school and scholar. To the ancient
Greeks, it was assumed that spare time… that time not devoted to working…
would be used for learning. This affinity for learning produced the first great
thinkers of mankind; such as Socrates, Plato and Aristotle… who famously
taught their students under a tree. "Leisure time" produced the greatest mathematicians, architects, sculptors, artists, musicians, poets, athletes and military
leaders that have ever lived.

Remarkably, for instance, Alexander the Great; perhaps the most brilliant
military leader of all time, actually took along with him on his conquests
naturalists to learn the geology and biology of the areas they were conquering.
He even had them send plant and animal specimens all the way back to Greece
for his teacher, Aristotle, to study.

In today's world, the omnipresence of technology increasingly isolates our
young people from the natural world. Yet, it is the natural world, the real world,
which inspires and provokes us to learn and to create. From Isaac Newton's
apple tree to Ben Franklin's kite-flying, many of the great inventors began their
journey of discovery with outdoor experiences.

Today's youth, with all their gadgets, you would think, should never get
bored. Whether at school, at church, or at home, everything in our culture
today seems to be geared toward their entertainment.   Yet, it seems to me that
they are more bored now than ever. Certainly, when I was growing up, I had

none of these electronic stimulants, yet I can scarcely remember ever being bored.

Just as the original meaning of the word for leisure has become disconnected from learning, the word "bored" has entered our lives. According to Patricia Spacks, professor of English at the University of Virginia and author of *Boredom: The Literary History of a State of Mind*, boredom is a relatively new human phenomenon. In fact "being bored", as we would say, used to be known as "committing acedia". Acedia was considered a sin, and was viewed as the willful devaluing of the world and its Creator; hence the old saying "idleness is the devil's workshop". Ideally, leisure time and study go together. Experiencing this type of leisure; once known as "scole", from whence our word for school is derived, is far less "boring".

Nature can be brought into the home to enhance our indoor time too. Indoor plants can improve our health, rest and relaxation. They help purify air in the home, and they bring nature's beauty into the residence. Fake plants are an atrocity to me…especially ones placed at graves in cemeteries, where they soon end up piled as garbage in some corner of the cemetery or thrown away. I consider them as litter…an example of the myriad of useless "stuff" that shouldn't even be manufactured in the first place. When I was growing up, my parents had a large fig tree in the living room and the spare bedroom was a veritable greenhouse, with all sorts of houseplants. This brought nature inside and created a cozy and relaxing atmosphere.

Watching Mr. Miyagi tend to his bonsai trees in the *Karate Kid* movie in 1984 instilled in me a fascination with bonsai. Like the Karate Kid, I was a skinny teenager at the time…fourteen and going through an awkward puberty…and could definitely identify with the movie. Several times, I had had to run home from school to get away from bullies in elementary school. By now, I had faced them and it was going pretty well…all it took was standing my ground and letting fly a few well-directed punches to the other kid's face. So, I watched the movie closely…hoping to learn some tips that might help in defending myself if and when the next bully started in on me.

Mr. Miyagi was awesome! He was a martial arts master, a bonsai master, and he even had cool classic cars! Yet, he was a gentle, short little "grandpa" character. In particular, his combination of "art" and "force" interested me…the same man that could expertly shape a tree into a peaceful emblem of nature inside the home…could kick ass to defend his home from outside forces. The movie gave me an appreciation not only for martial arts and bonsai, but the Japanese culture that was so tuned into both.

A few years ago, a famous bonsai tree attracted considerable media attention. Just five years after the Mayflower set sail in 1620, the "Hiroshima bonsai tree" was planted. This Japanese white pine, which was potted 393 years ago, was approximately two miles from where the atomic bomb was dropped by

American forces over 70 years ago. Somehow, the tree survived the explosion and the family that had cared for the tree for five generations gave it to the United States in 1975 in advance of the country's bicentennial. Its caretakers, however, wish to focus the attention on the tree's rule in peace rather than in war. The bonsai "was not given because of Hiroshima," says Kathleen Emerson-Dell, who helps care for the tree at the U.S. National Arboretum in Washington, D.C. "It was a gift of friendship, and connection—the connection of two different cultures."

The Arboretum wasn't even aware of the Hiroshima connection until 2001, when two grandchildren of the bonsai master Masaru Yamaki, who gave the tree; visited the National Bonsai and Penjing Museum at the arboretum; looking for their grandfather's tree. Yamaki's beautifully crafted little trees were protected in a walled nursery when one of the two bombs that is credited with ending World War II destroyed Hiroshima, killing approximately 140,000 people. Only a few feet tall, the stubby tree has a thick trunk and wires to help the branches keep their shape. (Becker, 2015)

Dwarfed trees occur naturally in the wild, where harsh growing conditions stunt their growth but do not affect their longevity. The steep, rocky outcrops on the Berea College Forest, for instance, feature small eastern redcedar, *Juniperus virginiana*; chinkapin oak, *Quercus muehlenbergii* and other species on these dry, rocky sites that are as old, or older, than the huge trees growing a couple hundred feet lower in elevation in fertile cove sites. The nearby Kentucky River palisades cliffs contain some of the oldest trees in the state, but they are not particularly large…due to slow growth.

Naturally occurring dwarf trees growing in the mountains of China were prized because of their unique aged and gnarled appearance. Taoists, thousands of years ago, believed that recreating such aspects of nature in miniature form in some way endowed that recreation with a certain magical concentrated energy. Known as Penjing in China, this art form means "tray scenery". Penzai or Punsai, translates into "tray plant"…and generally includes miniature rocks along with trees to create a tiny landscape.

The Chinese developed the pruning and binding techniques that create an aged appearance and unusually shaped features. Sometimes, trees appear to be purposely shaped to resemble dragons, serpents, or yoga positions. The first pictorial evidence of the Penzai tree appeared in 706 A.D. in the tomb of Prince Zhang Huai, when archeologists discovered painted frescoes depicting female servants carrying Penjing containing miniature trees and rocks.

During this same time, during the Hang Dynasty, Chinese monks migrated to Japan and other parts of Asia, taking Penzai and the art form with them. Japanese monks soon learned the techniques required for making the miniature trees, and it became known there as Bonsai. The Japanese developed methods of their own, and many believed the bonsai symbolized the harmony between man, soul and nature. In both countries, the trees gained honor and reverence. An ancient scroll, for instance, dated around 1195 A.D. expresses the appreciation and pleasure the author derived from gazing at the unusually formed little trees.

From the monasteries of Buddhist monks, bonsai trees spread into the homes of the royals and the affluent. The trees became status symbols and reflected honor. Bonsai trees were a highly regarded art form by the fourteenth century. Originally displayed the plants outdoors, the affluent later created special shelves indoors which provided the bonsai with a place of honor. Pruning techniques by the 16th century generally removed all but the essential parts of the tree; creating a minimalist effect. This reflected Japanese culture and philosophy, which believed that refining one's lifestyle meant eliminating all but the necessary. Modern Japanese gardens, even today, reflect these simplistic elements.

Bonsai trees gradually became more widely available in Japan to people of all social classes, and by the mid-1600's, the miniature trees began to be sold by merchants in other parts of the world. Provinces in Japan began holding bonsai exhibitions and competitions by the end of the 1700's, and in 1806 a bonsai was given to Queen Charlotte of England. The popularity of the trees also firmly engrained the bonsai into Japanese culture. Replicating the techniques of Taoists monks, the Japanese began creating a variety of miniature landscapes. Trees combined with buildings, people or rocks became known as bon-kei. Sai-kei was the term used for replicating a specific type or area of landscape.

News of the unique miniature potted trees spread to other parts of the world beginning in the mid-1600s, as merchants ventured to and from Asia. By the end of the 1700s, provinces in Japan began holding exhibitions and competitions. England's Queen Charlotte received a bonsai as a gift in 1806. Major world cities such as London, Paris and Vienna, eventually featured the trees in world exhibitions and the Japanese gradually started to share some of their secrets in the art of creating bonsai.

Berea Bonsai Studio and Studio and Nursery in Berea, KY is owned by Timothy J. Weckman, who has curated a public greenhouse for over 25 years and has a Master's Degree in Botany. He has studied bonsai with many notable teachers and is an apprentice of Warren A. Hill, Retired Curator of the National Bonsai and Penjing Museum in Washington, DC. Much of the information in this section comes from his website: http://bereabonsai.com/

After seeing Tim's bonsai trees, my wife and I dug up a nearly dead Japanese maple at the house that had been planted in a poor location by the previous property owners. We found it quite meaningful, as when we dug it up, we discovered that the tree was actually two trees which were intertwined. The roots were so intertwined that it took a lot of tedious work to separate them so that we could replant the two trees in an alternative fashion. We separated the two trees by a few inches so their trunks could develop as individuals; but kept the roots intertwined as much as this separation allowed. We envisioned the two trees as two people…us. "The two shall become one flesh" came to mind as we drew a comparison to the story of how the *Tree of Life* stands, somehow, on both sides of the *River of Life*.

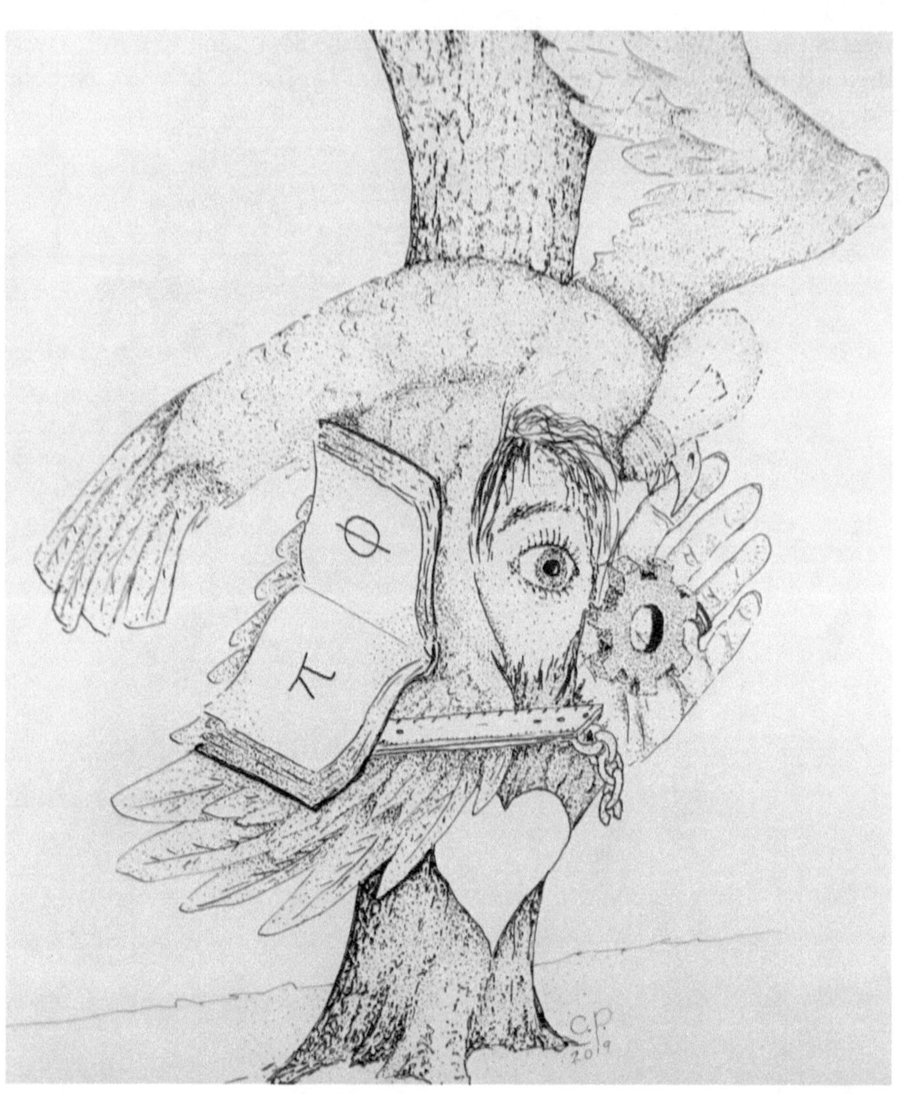

Chapter 8:
Awakening

"No man ever steps in the same river twice, for it's not the same river and he's not the same man." – Heraclitus

The Berea College Forest is a managed forested watershed, and its lakes provide some of the cleanest municipal water in Kentucky. Clean drinking water is often taken for granted until it is not available…and in many parts of the world it is a luxury. Some say that wars for water are the next flashpoint for world conflicts.

Water interests me almost as much as trees. I think children who missed out on playing in creeks, at beaches, or at least in mud puddles and ditches, missed out a great deal on growing up. Like bonsai trees and other plants that can be used to "set the mood" in our yards or even inside our homes, water features are also quite important for rest and relaxation. Humans seem to thrive on a savannah-like setting, near water. Perhaps it is genetic memory. Some innate instinct seems to bring us comfort to be in a setting that has the life-sustaining essentials…vegetation, shade, water…where we can see, hear and easily touch them.

I ran across a book about an Austrian forester, inventor and self-taught scientist named Viktor Schauberger. Schauberger was either a kook or a genius, and his ideas were pretty soundly denounced by other scientists of his day. Regrettably, both Walter Bitterlich (inventor of the Spiegel Relaskop) and Schauberger were around at the time of the rise of the National Socialists (Nazi) party in Germany. A Germanic country, Österreich ("east reich" or east nation) was annexed and came under Nazi rule. (The Sound of Music (1965) movie follows the Van Trapp family's experience during this time). Bitterlich served in the Wehrmacht and Hitler took an interest in Schauberger's ideas. An internet search of Schauberger's name will reveal quite a few "far out" stories; mostly focused on the supposed invention of the German "flying saucer", the BMW Flugelrad (flying wheel).

Did a forester invent a flying saucer? Conspiracy theories aside, the book *Living Water: Viktor Schauberger and the Secrets of Natural Energy* by Olaf Alexandersson, was a trip to read. The book tells the story of this Austrian forester, naturalist and iconoclast who opened the way to a completely new understanding of the vast potential of natural energy. Studying fish in streams and by closely observing the natural water cycle, Viktor Schauberger (1885-1958) was able to offer solutions to problems of energy transformation. He urged modern man to recognize that we were destroying the earth and sabotaging our own cultures by working against Nature.

Schauberger sounds a lot like the great inventor of our modern era Nikola Tesla when he describes water as the source of "free energy". Like Goethe, Schauberger mocked the narrowmindedness of conventional science and took issue with Newton's assessments; saying "I think it would have been much better if Newton had contemplated how the apple got up there in the first place!" He proclaimed:

*"The revelation of the secret of water will put an end to all manner of specula-
tion or expediency and their excrescences to which belong war, hatred, impatience
and discord of every kind. The thorough study of water therefore signifies the end of
monopolies, the end of all domination in the truest sense of the word and the start of
a socialism arising from the development of individualism in its most perfect form."*

Schauberger's ingenuity first became apparent when he, as a forester, saw a
fish effortlessly swimming in a fast-moving mountain stream and contemplated
that the fish was not swimming at all, it was "being swum" by the water. He set
out to understand how swirling water vortexes behaved. At that time, foresters
in the mountains where he worked…as in other mountain regions of the
world…relied on log flues to transport logs down the mountains. Ever go on a
"log ride" at a water park? You get the idea.

Viktor employed his water vortex ideas to log flue design. Scientists had
already "proven" the limits of weight and volume that could be transported by
water. Yet, Schauberger, a forester with no formal scientific education or
training, successfully produced a flue design that could transport far more
weight than was thought to be possible by scientists. Kook or not, his flue
worked and he began to look for further applications for his "water vortex"
concepts.

Schauberger had a clear vision of how fertility and health could be restored
to the earth. He joked, himself, that perhaps people were right that he was
deranged; yet he developed a number of ingenious machines which would
revolutionize farming, horticulture, forestry and aircraft propulsion. He invented water purification systems and showed how air and water could be harnessed
as fuels for machines.

Jules Verne, in *The Mysterious Island*, in 1874, had already imagined the
"burning of water" in both its broken down components of hydrogen and
oxygen, as replacing the use of coal in the future. However, Schauberger
imagined a whole new technology that required no explosion or burning. It
used what he called "implosive" energy and "diamagnetism"…most of which
have yet to be developed. But, like Werner von Braun…the famous German
rocket scientist who run the American space program….Schauberger was also
brought to the United States and tapped for his ideas. Some speculate they have
been or are being further developed. His "flying saucer" invention may not
seem so incredible if you think about the full extent of Von Braun's work.

Von Braun's plethora of ideas and plans have still not been fully realized.
The space station featured in *2001: A Space Odyssey* by Stanley Kubrick (1968)
was based on his actual design. The NASA space shuttle was his design for a
reusable spacecraft that could be used to transport cargo to multiple larger spacecraft in Earth orbit. He was already working on interstellar space travel before
he was used by the Nazis for war purposes. He was arrested by the Gestapo for
continuing to work on space travel instead of missiles for the war. After the war,

he published Das Marsprojekt (1952); technical specifications of his plan for a massive manned scientific expedition to Mars that would involve ten spacecraft with 70 crew members. He foresaw this coming to fruition in 1968. Von Braun's plans are still influential today, and while they may seem far-fetched or even megalomaniacal, they seem strangely prophetic as well. For instance, he envisioned the colonization of Mars and establishment of a government there…and even provided the title for the leader of the parliament of Mars: "Elon". Does that sound familiar? Who is heading up a Mars expedition now, with his rise from PayPal and Tesla inventor to SpaceX? Well, of course, that would be Elon Musk.

I wonder where we would be at today, in terms of scientific advancements and civilization, if the genius of scientists like Von Braun and Einstein, had been used only for beneficial purposes rather than for war? And, what if the ideas and inventions of scientists like Schauberger and Tesla had been wholeheartedly embraced, rather than abandoned? As technological advances continue, not only is the welfare of Mankind and Civilization at stake, but Life itself. Man is not only at war with himself…but with Nature itself. This was the primary concern of Viktor Schauberger who said:

"This civilization is the work of man, who high-handedly and ignorant of the true workings of Nature, has created a world without meaning or foundation, which now threatens to destroy him, for through his behavior and his activities, he, who should be her master, has disturbed Nature's inherent unity."

Schauberger understood that Nature has an inherent unity, and that Man's role as its "master" is not to be "high-handed and ignorant" of this. Man's true role as master is to create a world with meaning and foundation in unity with nature…not to dominate it as a selfish tyrant. For, according to Schauberger, the "highest wisdom" comes not from advanced learning and scientific discovery, but is understood simply by heart:

"The majority believes that everything hard to comprehend must be very profound. This is incorrect. What is hard to understand is what is immature, unclear and often false. The highest wisdom is simple and passes through the brain directly into the heart."

———————

"Have a heart" is what we say when we implore someone to care about something or someone. We don't say "Have a brain". And yet, the things of the heart are not taught in education anymore. Is it any wonder then that depression, suicide, and drug addiction are running rampant and decimating so many individuals, families and communities? How can we improve our natural world and wrest it from destruction from uncaring actions if we can't first help people to care in the first place? What will they care with?

The scariest movie I ever saw as a child was *The Shining* (1980). I think it was the first time I saw insanity portrayed…at least so realistically (supposedly) by Jack Nicholson…one my favorite actors. Perhaps the most famous scene is when Jack Nicholson's character busts through the door with an axe and says "Here's Johnny!" Imitating the introduction of Johnny Carson on the The Tonight Show (then starring Johnny Carson) it was such a great line to contrast horror with humor. I loved staying up to watch Johnny Carson.

I could see the humor in the "Here's Johnny" scene…even with the axe. But, the "typewriter scene" was particularly disturbing to me. A convincingly crazy Jack Nicholson is sitting at a typewriter obsessively typing:

"…all work and no play makes jack a dull boy all work and no play makes jack a dull boy all work and no play makes jack a dull boy all work and no play makes jack a dull boy all work and no play makes jack a dull boy all work and no play makes jack a dull boy all work and no play makes jack a dull boy all…"

Then, he went on a murderous rampage. "All work and no play" happens off the television screen too…with violent consequences. Keeping Johnny working, living off his labor, and giving him little time to play, has been the *modus operandi* of rulers since the beginning of time. What time Johnny does get, is filled with "bread and circuses", 24/7 television, video games, and media bombardment. This keeps his mind distracted, his heart tuned out, and he never actually has any free time. Trying to control who Johnny attacks when he finally snaps, to control the outcome, is also the strategy of tyrants. Orwell's books describe this process well.

Karl Marx, yet another German, has perhaps had a more profound influence on this world than anyone I have cited. Born in 1818, Marx was a philosopher, economist, journalist and revolutionary socialist. He studied political economy and Hegelian philosophy and collaborated with Friedrich Engels. Mostly, he is known for his two major publications: the pamphlet, *The Communist Manifesto* and the three volume work *Das Kapital*.

Collectively understood as Marxism, Marx's theories about society, economics and politics are explained as a conflict between the ruling classes and the workers. Marx predicted that, like other socioeconomic systems, capitalism produced internal tensions which would lead to its self-destruction. He predicted it would be replaced with a new system which would…he thought…bring about socio-economic emancipation.

Marx actively advocated for the implementation of his ideas; arguing that the working class should carry out organized revolutionary action to topple capitalism and bring about socio-economic emancipation. His ideas have found the support of many, as they sound pretty good on the surface. But, millions upon millions were murdered in Russia, China, and other places when governments killed their own people to force a Marxist system of "emancipation" upon them.

The Gulag Archipelago, written between by Alexander Solzhenitsyn about his own experiences as a prisoner in the Soviet Union's forced labor camp system, is a heart-wrenching book that I think should be more widely read. Solzhenitsyn concluded, after witnessing and enduring horrific atrocities under communism, that it's not about one system over another…one philosophy or nationality or religion over another. His conclusion was this:

"If only it were all so simple! If only there were evil people somewhere insidiously committing evil deeds, and it were necessary only to separate them from the rest of us and destroy them. But the line dividing good and evil cuts through the heart of every human being. And who is willing to destroy a piece of his own heart?"

"Gradually it was disclosed to me that the line separating good and evil passes not through states, nor between classes, nor between political parties either -- but right through every human heart -- and through all human hearts. This line shifts. Inside us, it oscillates with the years. And even within hearts overwhelmed by evil, one small bridgehead of good is retained. And even in the best of all hearts, there remains … an unuprooted small corner of evil."

"Since then I have come to understand the truth of all the religions of the world: They struggle with the evil inside a human being (inside every human being). It is impossible to expel evil from the world in its entirety, but it is possible to constrict it within each person."

Solzhenitsyn wrote of his experiences so others could learn of the horrors brought to his people by the Soviet government…but not only to learn of it with their minds, but absorb it with their hearts so others would not go down this same path experience it too. Such horrors can come not only from a forced Communist system, but a forced religious system or any other system where the "evil that cuts through every heart" is ignored. He said, in a quote not attributed to his book:

"The sole substitute for an experience we have not ourselves lived through is art and literature."

Thus, the more people can be exposed to art and literature, the more of this substitute for experience can be obtained. Is it any wonder, then, that the intellectuals…the educated…are often the first to be killed when a totalitarian regime comes to power? Is it any wonder that books are burned and free speech impaired…and even art is put under control of the government?

Yuri Bezmenov, KGB defector, escaped to the West in 1970 and spent the 1980's warning how the Marxist-Leninist system of psychological warfare and ideological subversion how works. According to Bezmenov, it is a four step process beginning with demoralization. Working primarily through the American education system, he detailed how this process virtually renders the generation growing up under it unable to assess truth no matter what facts are presented. This process, he attested, was completed by 1985. The next step is destabilization and works through economic and military subversion. The wave of outsourcing of our industrial and military capabilities took care of that next over only a few years. The third step is crisis. He describes how the many leftist coups in Central America were examples of this…but in America the hope was to install a Big Brother Government through our political system; citing Mondale then as a perfect vehicle for this. Finally, the last stage is "normalization". He explains that when the tanks rolled into Czechoslavakia in 1968 and a Marxist-Leninist puppet government was installed, the announcement was made that the situation had been "normalized" and this is what awaits every nation at the end of this process. My own personal observation is that the Marxist-Leninist strategy described by Bezmenov continued unabated after the so-called fall of the Soviet Union and is now almost complete as we are exposed constantly to all manner of crisis designed to get us to the fourth and final normalization stage.

Today, we can "google" anything and read anything we want. We can still access things like Bezmenov's interviews on YouTube. But, what difference does it make if our society is already demoralized, our communities destabilized, and our daily news cycle filled with crises? All manner of information is readily available, but are we really doing anything with it? How many people…particularly younger people who have grown up with cell phones…regularly read whole books? How many go to art museums? Is art even taught in school anymore or essays written about classic books? Are matters of the "heart" even allowed? It seems that even meaningful conversations are becoming almost impossible these days because of "political correctness". Censorship can come much more subtly than with burning books.

This is why, today, I believe it is more important than ever for young people to spend time in nature. Nature is not censored. Nature still speaks to the heart. The Nordic countries; Norway, Sweden, Denmark, Finland, and Iceland…always seem to come in at the top of the "happiest countries" lists that come out each year, along with New Zealand, Australia, and Canada. Why? Some would say it is because their society is ran by what most would describe as "democratic socialism". But, they also have a favorable business climate and economic freedoms associated with capitalism. I think their happiness has something to do with their ability to achieve this balance.

How do they achieve this balance? I think they have a work/life balance that is rooted in exposure to free time in nature. All these countries have great access to nature, are mostly uncrowded, and the Nordic countries even have the "right to roam". This right to roam the forests and waterways, pick berries and mushrooms, and enjoy the landscape virtually anywhere, seems to promote a culture which naturally is conducive to providing other common benefits like universal healthcare, a living wage, free education and so on. It also instills the work ethic, physical and mental health, and stable economy needed to support it.

Nature and Freedom seem to be the essential ingredients for the heart and mind to survive intact and thrive to produce things of both beauty and utility.

Capital…Labor…Resources…these things are major topics that various sciences and politics have come up with systems to try to run humanity in a better direction with regard to them. Utopian societies, "hippie" communes and the Amish communities set up and live out various versions of their own ideas based on biblical and/or philosophical ideas. Some success is usually found for a while; then conflict arises and these communities are split up or dissolved. I know former Amish who came out of their communities as well as self-proclaimed "hippies" who left their communal living places over disagreements and conflicts. So long as people are free to come and go, to create and dissolve communities, to think and speak and do…so be it. When people are free, they have the opportunity to follow their dreams, to follow their hearts…to find new and better solutions to try. Some just might work and can then be emulated. If they don't work, the people naturally move on and try something else! No coercion. No one had to die to enforce someone's "ism". I am always drawn to the people who at least try something counter to the status quo…people who experiment and explore and wonder.

So many problems besiege us every day in our modern world of mass media, and it is difficult to rest…to truly rest from real and imaginary problems that confront or threaten to confront us. Nature is a favorite respite from the hectic life, the rushed barrage of information coming at us and from the noise of civilized life. I am fortunate to work among nature most days; where I can experience nature on an almost daily basis. But, even so, the demands and pressures of life get me frazzled from time to time. I have learned that I must make time for destressing or suffer the consequences.

When faced with doctors urging me to get back surgery for a herniated disc, and again urging me to get on prescription blood pressure medications, I discovered instead a guy nicknamed "The Iceman". Wim Hof, aka *The Iceman*, holds 26 world records; including one for longest ice bath…which he does for well over an hour on stage for audiences and doesn't even shiver…his body temperature remaining normal. The guy climbs frigid mountains in record time while wearing only shorts, and completed a full marathon (42.195 kilometers (26.219 mi) above the Arctic circle, in temperatures close to –20 °C (–4 °F) while wearing only shorts.

Hof uses breathing techniques based on Tibetan *Tummo* meditation and controlled hyperventilation to achieve his feats and for various health benefits as well. He also does yoga and stays insanely active. Hof has subjected himself to all manner of scientific tests and proven that what scientists have said is humanly possible can be done. He even believes it also can keep people from ever…ever…becoming sick. Scientists, at first, thought Hof was a fraud until they studied him. Then, they thought he was a freak of nature…until he trained others to do what he can do.

I'm not really interested in swimming under ice or doing a marathon in Antarctica or Death Valley like Wim does with seemingly super-human ease. But, I was interested in avoiding back surgery and being put on blood pressure medications. Hof's techniques and prayer helped me overcome both with no medications. Scientists viewed his feats as impossible…just like scientists said Viktor Schauberger couldn't float a log over a certain weight on a certain volume of water. Yet, there is always more to add to whatever body of information "science" holds up to us as settled.

When phenomena are not understood…yet happen anyway…some might call it "magic". But magic is simply science not yet fully understood. According to Wim Hof, his abilities are attributed to "mind over matter". But what is "mind"? According to Max Planck, it is everything. Nature still has a lot of secrets science has yet to discover. We should not be so quick to discount and say anything is "impossible"…including matters of the mind and human conscience.

I am a forester by profession and an artist by heart, and I hope that I have been able to convey some ideas that you might "take to heart" in this book. I had intended to call it "The Art of Forestry" when I began writing it. Forestry is indeed both an art and a science. The development and application of forestry as such, requires not only use of the brain of the heart. Aristotle considered the heart to be the center of reason, thought, and emotion; higher than the brain in importance. Abu Nasr al-Barabi, ninth century Arabic philosopher, believed that the heart was the ruling organ of the human body. Auguste Compte, 19th century French philosopher, pronounced the brain to be a servant of the heart. If we are to live through this Anthropocene era, and come out on the other side with any semblance of nature intact, we must find a new harmony with each other and with nature. This will take not only employing both sides of our brain (science and art) but also our heart. Inside the heart of man, it has been said, is the spirit.

"I used to think the top environmental problems were biodiversity loss, ecosystem collapse and climate change. I thought that with 30 years of good science we could address those problems. But I was wrong. The top environmental problems area selfishness, greed and apathy…and to deal with those we need a spiritual and cultural transformation -and we scientists don't know how to do that." – Gus Speth

Gus Speth's credentials, accomplishments and awards during his life's work on environmental issues would take a full page just to list. He served at the top of academia and on the highest level tasks forces and committees on the environment for decades. And yet, he comes to the conclusion that science doesn't know how to do what is needed to combat our environmental problems. What is needed, he realizes, is a "spiritual and cultural transformation".

Many perceive religion as part of the problem, not the solution. If we talk in terms of spirituality, it has to be vague enough to sound "new age" or we're automatically discounted as "religious"…a term which is becoming increasingly equated with being "anti-science" and thus not worthy of consideration. Many cite the Book of Genesis where God told Adam and Eve to "subdue" the Earth and to "have dominion" over every living creature as proof that "religion" is to blame. These English words, translated incorrectly I think from the original Hebrew, carry negative connotations to most people. It sounds like we are given marching orders to "dominate" and "take over" to "rule over". Far too often, people have acted this way toward Nature and each other. That's the carnal human nature. It comes easy.

Viktor Schauberger knew this wasn't right. To be master over Nature is to be a co-creator with her, not a destroyer of her. He knew we must humble ourselves before Nature's innate order to obtain Her Wisdom. This is *Gottesblick* in action…*Order* that displays the *Highest Virtue*.

One suggested translation of the word commonly translated as "dominate" is a similar Hebrew word which means "to lower one's self". This word would fit Schauberger's views on the Man's proper relationship as master over Nature. If so, that would provide an almost opposite interpretation…meaning essentially that we are to humble ourselves before Creation and put her first to be her partner in creation.

Even the English word dominion is used, this not necessarily as bad as some would have us believe. The root word for dominion, after all, comes from the Latin *dominio*; from *dominus* "lord", from *domus* "home". So, the original word had to do with making ourselves a home. Do you trash, pollute, and destroy your own home? Well, then we shouldn't do that to Nature either. Everyone is thus the lord or ruler over their own home. As ruler over our home, then, what do we do? Do we tear it apart, burn it down, or neglect it? No. We protect it and maintain it. We even personalize it, decorate it to look nice and feel comfortable for ourself, our family, and our guests…do we not?

Of course this is what makes sense when you think about how Christ lived and how He instructed us to love one another and to serve one another. Let's consider also the following Old Testament instructions concerning the land:

"But the seventh year thou shalt let it rest and lie still; that the poor of thy people may eat: and what they leave the beasts of the field shall eat. In like manner thou shalt deal with thy vineyard, and with thy oliveyard." -Exodus 23:11 (KJV)

I have heard a lot of pastors preach about how we need to give ten percent of our money to the church…I've heard a lot about that tithe. But, I don't remember ever hearing a pastor instruct his congregation to leave their land fallow every seven years so that poor people and wildlife can live off of it and to give the land rest. Having read this at an early age, I was actually surprised to find that farmers didn't do this. Sounds pretty sustainable…like conservation and a social program wrapped up on one effort.

So, does God really care if we do that? You bet…big time:

"Speak to the sons of Israel and say to them, 'When you come into the land which I shall give you, then the land shall have a Sabbath to the LORD. Six years you shall sow your field, and six years you shall prune your vineyard and gather in its crop, but during the seventh year the land shall have a Sabbath rest, a Sabbath to the LORD; you shall not sow your field nor prune your vineyard… You, however, I will scatter among the nations and will draw out a sword after you, as your land becomes desolate and your cities become waste. Then the land will enjoy its Sabbaths all the days of the desolation, while you are in your enemies' land; then the land will rest and enjoy its Sabbaths. All the days of its desolation it will observe the rest which it did not observe on your Sabbaths, while you were living on it." - Lev. 25:2-4; 26:33-35

So, because the people didn't let the land rest, God forcibly removed the people from the land so that the missed Sabbaths could be restored. Wow! I'd say He is pretty serious then about the environment and taking care of the land…so it could "enjoy its Sabbaths"…to rest. So, evidently our interpretation of "subdue" and "to have dominion" does not fit the true intention. Farming and ranching used to be referred to as "husbandry" and was akin to a man taking care of his wife. Husbandry of the land is thus supposed to be modeled on the example of a loving marriage; not slavery. It is without rape and subjugation. Neither the land, nor women, should ever be abused or mistreated.

The biblical model for a husband is for him to be willing to lay down his life for his wife as Christ laid down his life for the Church; which is called the Bride of Christ. In both instances, the husbandman and the husband, Man is to do everything in love and take care of both the land and his wife. Both produce and nurture life. Those who come to steal, kill and destroy…whether in the name of religion or not…follow the pattern of the "thief" who is the opposite of the "good shepherd". (John 10)

One of the earliest female contributors to scientific study, art, music composition, and writing was the great medieval Christian mystic, pharmacist, theologian and proto-feminist Hildegard of Bingen. One of only four "doctors of the church", Saint Hildegard was canonized and named a Doctor of the Church by Pope Benedict XVI in 2012.

Considered the founder of scientific natural history in Germany, Hildegard was known as the "Sibyl of the Rhine". Born into a noble family in 1098, she became a Benedictine nun at the Monastery of Saint Disibodenberg at the age of 18. Hildegard reportedly had visions since the age of three, and wrote them down in her *Scivias* (Know the Ways) over a ten year period, in middle age. Pope

Eugene III read it, and in 1147, encouraged her to continue her writing. She went on to produce short works on medicine and physiology, and hundreds of letters of correspondence between many prominent leaders of the time… including Kings, Queens and Popes.

Hildegard's visions enabled her to see humans as "living sparks" of God's love, and she saw the harmony of God's creation and the place of humanity in that…a unity that was not apparent to many of her contemporaries. She often boldly spoke out against what she saw as a spineless church and evil state fighting over prestige and power.

Hildegard left behind an astounding body of work. She composed over 80 songs, wrote *Causae et Curae* and *Physica*, which offered nutritional and medical advice and included thousands of herbal remedies…even discussing sexuality in surprisingly modern terms. She went on speaking tours in her later years, where she denounced corruption of the church, and even invented her own language and alphabet.

Hundreds of years after *Scivias*, the great statesman, artist, playwright and scientist Johann Wolfgang von Goethe, upon seeing it, wrote: "an old manuscript containing the visions of Saint Hildegard, is extraordinary". Carl Jung, founder of analytical psychology, studied Hildegard's works as well, and her mandala images would be a reference point for his process of individuation.

The original *Scivias* manuscript was lost in the chaos of World War II, when Dresden came under the occupation of Soviet troops. The manuscript remains missing to this day. Fortunately, however, in 1925 photographs of the original *Scivias* were taken as part of a series of exhibitions in Cologne. Additionally, in 1933, a duplicate manuscript was created and safely kept at the Abbey of St. Hildegard, where it remains today.

Religious leaders through the ages have discussed and fussed about how to interpret the various books of the Bible and which ancient texts to include among the compilation we refer to as "The Bible". Even after the discovery of the Dead Sea Scrolls in 1947 proved the survival of and authenticity of the scripts; it also opened a "can of worms" when fragments of other scripts like the Book of Enoch were found among the caves as well. The Book of Enoch is pretty interesting…it is a continuation of what happened to Enoch after he was "translated" and "not found". It sounds a lot like Enoch took a trip in a spaceship! New Agers might say he became enlightened; activated his star tetrahedron, and traveled to a higher dimension. Alien enthusiast might imagine he was abducted. Similar accounts can be found describing Elijah as being taken away in a "chariot of fire" (II Kings 2:11) and what seems to be an aircraft described by Ezekiel as concentric, spinning, sparkling flying "wheels". *(Ezekiel 1:16)* Ancient Sanskrit epics and Hindu texts describe the same things, in great detail. Various types of flying machines called *vimanas* sound much like what we would think of today as spaceships.

Maybe I watched too much *Battlestar Gallactica* as a child. But, even people like Elon Musk are publicly saying things like: "There's a billion to one chance we are living in base reality". Meanwhile, famed physicist Michio Kaku speculates: "Our grandkids will lead the lives of the gods of mythology. Zeus could think and move objects around. We'll have that power. Venus had a perfect, timeless body. We'll have that too. Pegasus was a flying horse. We'll be able to modify life in the future."

Bob Warren, former forestry technician at Berea College, had a doctorate in molecular biology. He studied the genes of flies in Wisconsin and centipedes in Australia and could tell you how genes of insects can turn off and on…causing a fly to grow or not grow wings. What if there are genes in life forms that we don't even know can "turn on" and produce a version of them we have never seen before? Looking up at the night sky and its multitude of stars…or peering into a microscope at a seemingly infinite space in minutia…the vastness of Creation is seemingly endless. What an adventure it is to explore what is here!

Appalachia is the most biologically diverse temperate region in the world. I don't have to look through a microscope or telescope to see wonder upon wonder. A short hike can take me into the midst of a vast diversity of life here that can provide a lifetime of opportunities to discover. Maintaining a sense of wonder and a zest for learning naturally will cause us to ask questions. Never stop asking questions…and be sure to question others' answers too.

I like what renowned physicist, Richard Feynman, once said: *"I would rather have questions that can't be answered than answers that can't be questioned."*

Berea College's historic Mission is to "Promote the Cause of Christ". Berea College, and then the town of Berea, was named for the place in the Bible called Berea. They questioned the Bible and the Constitution and decided that slavery was wrong and that all men and women were created equal and deserved to be treated that way. They decided to act on it. They found their inspiration and the name for their new College in the Book of Acts.

"And the brethren immediately sent away Paul and Silas by night unto Berea: who coming thither went into the synagogue of the Jews. These were more noble than those in Thessalonica, in that they received the word with all readiness of mind, and searched the scriptures daily, whether those things were so. Therefore many of them believed; also of honorable women which were Greeks, and of men, not a few."
–Acts 17: 10-12 KJV

"Readiness of mind", you might say, could be interpreted as "open-minded". Certainly, I think being open-minded is key to harmony. The Greeks had readiness of mind and listened to Paul and Timothy, while they met with disputes and beatings elsewhere. Authorities and church leaders were constantly "stirring up" opposition against them. The Bereans Paul and Silas addressed were Jews…some believing Paul that Jesus was the Messiah was prophesied; such as in *Isaiah 53:*

"Who hath believed our report? and whom is the arm of the LORD revealed?… He is despised and rejected of men; a man of sorrows, and acquainted with grief: and we hid as it were our faces from Him; He was despised and we esteemed Him not. Surely He hath borne our griefs, and carried our sorrows: yet we did esteem Him stricken, smitten of God and afflicted. But He was wounded for our transgressions, He was bruised for our iniquities: the chastisement of our peace was upon Him; and with His stripes we are healed. All we like sheep have gone astray; we have turned every one to his own way; and the LORD hath laid on Him the iniquity of us all… He hath poured out His soul unto death: and He was numbered with the transgressors; and He bare the sin of many, and made intercession for the transgressors."

Not all of the Bereans believed Paul's message of Christ…but they all had "readiness of mind"…they were open-minded to hear it. Berea College, likewise, is not made up of only Christians…far from it. Berea is not a "Bible college" and has no denominational affiliation. Faculty and Staff are not required to be Christian. Yet, the purpose of the College remains to *"Promote the Cause of Christ"*.

So, what IS the "Cause of Christ"?

First, let's consider who is Christ? He is called many things: *The Lamb, The Lion of the Tribe of Judah, The Door, The Way, The Truth,* and *The Life.* He is referred to as *Living Water* and *The Light of the World.* He is also called *The Word:*

"In the beginning was the Word, and the Word was with God, and the Word was God." John 1:1

The reference to "the word" is important. The Greek word is "logos" and it means "idea" or "purpose". God's purpose therefore was centered in Christ:

"And the Word was made flesh, and dwelt among us, (and we beheld his glory, the glory as of the only begotten of the Father,) full of grace and truth." Jesus…the name means "God saves"-Matt 1:21…was born a man. He lived a life in which he was tempted to sin like all other men — yet unlike all other men he had the power to resist the temptations of the flesh. He fulfilled God's will perfectly so that he was "the image of the invisible God" -Col 1:15

The name "Jesus" is derived from a form of the Greek word *"Jeshua,"* meaning "Jehovah-Savior" or "the Lord saves." The title "Christ" comes from the Greek word *"Cristos"* and means "anointed one" or "chosen one". Thus, Jesus is the anointed, chosen one to "save" Mankind…to reconcile us back to God. He became our sin so that we could become his righteousness…and thus obtain eternal life. And what was Christ's message? It can be "boiled down" to this:

"…love one another as I have loved you…by this shall all men know that ye are

my disciples, if ye have love to another." – Jesus, in John 13:34

Elsewhere, we read that "God is Love" and are instructed to live up to being created "in the image of God", like Jesus, by John the Apostle:

"Beloved, let us love one another: for love is of God; and every one that loveth is born of God, and knoweth God. He that loveth not knoweth not God; for God is love. In this was manifested the love of God toward us, because that God sent His only begotten Son into the world, that we might live through Him. Herein is love, not that we loved God, but that He loved us, and sent His Son to be the propitiation for our sins. Beloved, if God so loved us, we ought also to love one another." –I John 4:7-11

Not all the Bereans believed, and not everyone does today. Even among believers, there was, and still is, much disagreement about certain things…just like there is among any other group of people. The idea of people loving one another, despite their differences, motivated a Christian pastor to found Berea College. John Fee and his supporters believed we are all "created of one blood".

"There is neither Jew nor Greek, there is neither bond nor free, there is neither male nor female: for ye are all one in Christ Jesus." – Galatians 3:28

Within the Church, the idea of being "One in Christ" and members of "One Body" is that each person has their own unique value and role to play.

"For as in one body we have many members, and the members do not all have the same function, so we, though many, are one body in Christ, and individually members one of another. Having gifts that differ according to the grace given us, let us use them: if prophecy, in proportion to our faith; if service, in our serving; the one who teaches, in his teaching; the one who exhorts, in his exhortation; the one who contributes, in his generosity; the one who leads, with zeal; the one who does acts of mercy, with cheerfulness." – Romans 12:4-8

The idea of "Oneness"…is not limited to spiritual concepts, but rather is foundational to holistic thinking with regard to science, biology, and ecology. While Aldo Leopold admonished us to "Think Like a Mountain", another man helped us to "think like an ocean" to save the great fisheries of the world.

Ed Ricketts wrote in 1936: *"Inter-relation seems to be pretty much the keynote of modern holistic concepts, wherein the whole consists of the animal or the community in its environment, the notion of relation being significant."* (Unpublished Zoological Introduction to Between Pacific Tides; from the WesternFlyer.org)

In the spring of 1940, writer John Steinbeck and his close friend, marine biologist Edward F. Ricketts, chartered the Western Flyer, with Tony Berry as captain, to take them on a scientific expedition to the Sea of Cortez. For 6 weeks in March and April, they crew collected marine invertebrates in this remote body of water. Studying and cataloguing communities of marine invertebrates had been Ricketts's life work, resulting in his seminal text on Pacific intertidal communities, Between Pacific Tides, published in 1939 and still in print. Ricketts's work was and is significant because his is an ecological model, noting connections between animals and communities in particular environments, rather than focusing on scientific classification.

The 1940 trip to Baja resulted in Steinbeck and Ricketts's publication of Sea of Cortez in December, 1941 (reissued in 1951 as the Log from the Sea of Cortez, which cut the phyletic catalogue of invertebrates and added an essay by Steinbeck, "About Ed Ricketts"). Steinbeck claimed that Sea of Cortez was his favorite among all of his works. And one critic remarked that there is more of Steinbeck, the man, in this work than in any other.

This boat honors the spirit of Steinbeck and Ricketts's trip, their shared ecological perspective, and their curiosity about the natural world. This boat is the actual vessel that took them on their historic voyage. This boat is a pioneering vessel in the history of ecology. Western Flyer, the Purse Seiner that author John Steinbeck and marine biologist Edward F. Ricketts chartered to the Sea of Cortez in 1940. The goal of all Western Flyer education programs is to integrate environmental studies, history, art, and humanities—a holistic approach.

The most intriguing and exciting project I have had the opportunity to work with while serving as the Berea College Forester is our involvement with the Western Flyer boat restoration project. Supplying clear, eighteen foot white oak logs for the restoration of a historic ship would be cool in itself, but to be a part of an educational program which will use this ship to teach community place-based, holistic understanding of our planet is especially satisfying and fitting. Ben and Sam, our contract horse and mule loggers; also known as biological woodsmen, got the logs out. Want to see the logs being harvested from the forest? Search "Western Flyer Berea" on You-tube.

"'Let us go,' we said, 'into the Sea of Cortez, realizing that we become forever a part of it; that our rubber boots slogging through a flat of eel-grass, that the rocks we turn over in a tide pool, make us truly and permanently a factor in the ecology of the region. We shall take something away from it, but we shall leave something too.' And if we seem a small factor in a huge pattern, nevertheless it is of relative importance. We take a tiny colony of soft corals from a rock in a little water world. And that isn't terribly important to the tide pool. Fifty miles away the Japanese shrimp boats are dredging with overlapping scoops, bringing up tons of shrimps, rapidly destroying the species so that it may never come back, and with the species destroying the ecological balance of the whole region. That isn't very important in the world. And thousands of miles away the great bombs are falling and the stars are not moved thereby. None of it is important or all of it is." (WesternFlyer.org)

Sky blue tree marking paint sprayed out of my paint can onto the massive white ash. My Biltmore stick was barely long enough to record its diameter. The timber sale would be cut "on shares" by horse and mule loggers. It was an emotional week for me. Shearer Hollow's topography was up and down, with steep slopes. Many of the big trees I was marking were biologically mature and I felt ok about conducting a harvest here…but it was sad to see the demise of the ash trees. At the same time, the realization that I may be seeing big mature ash trees for the last time, gave the experience a sense of something to be treasured. I imagined foresters marking dying American chestnut trees a hundred years ago and how now it is exciting to find one large enough to produce nuts. Emerald ash borer was here and the trees were doomed. Most were already showing signs of decline. Looking up through the sparse canopy of every great ash I marked, I saw one that looked like a woman with her arms thrown up in despair…the leaves were all gone from her hands, but a shock of leaves was stilling hanging on to the top of the middle of the canopy…her hair.

Marking timber, I normally work along the contour of the hills. However, I was especially getting a workout this week because I chose to work the steep hills in an up and down pattern. I had just met the woman, who is now my wife, online. We were getting to know each other through texting, and I only got a cell phone signal at the top of the ridge. It took about an hour to work my way to the bottom and back up to the top. I took a lot of photos of the trees I was seeing, and some selfies. About three resends later, most would go through. She texted me a photo of herself taking a break under her favorite tree at the Mountain Home Veteran's Hospital where she worked…a massive pin oak near the parking lot. A while later, I received a photo of the biggest catalpa tree in Tennessee…a short walk on their campus. I couldn't wait until the weekend.

"I have to get this special woman something special…make a good first impression", I thought, as I wandered into the *PeaceCraft* store in Berea the next day. It was uncanny we had connected so quickly, after only one week of texting. The previous weekend, her profile caught my eye. With her long, black hair and dark almond eyes, she looked Native American. I couldn't resist but to join the site so I could contact her. When I saw her idea for a "perfect date" was to "go hiking in a historic forest, and then go get a beer and tacos", I was smitten.

I was going to make the three and a half hour drive to east Tennessee to meet her the next day. Turns out, she wasn't Native American; she was one-quarter African…the rest being Italian and Irish. I bought a bag of organic Ethiopian coffee and a hippie-looking ring. As I was about to check out, the lady at the counter called my attention to some hand-made quill cards close by. My eye was drawn to one with a lotus flower on it. I picked it up and saw that it had small letters at the bottom that spelled Namaste. I had no idea what

Namaste meant. But, I knew the lotus symbolized new beginnings…beauty grown out of the mud…and that it was important in ancient Hinduism, Buddhism, and Egyptian culture. A symbol of rebirth and purity, it is associated with the idea of spiritual enlightenment. The unopened flower bud is representative of a folded soul that has the ability to unfold and open itself up to the divine truth. Like everything in nature…from the nautilus to the galaxies, to the pine cone…the flower petals open in a Fibonacci spiral. (https://www.lotusflowermeaning.net/)

I bought the card. Inside the card, I wrote a Bible verse:

"Behold, I will do a new thing; now it shall spring forth; shall ye not know it? I will even make a way in the wilderness, and rivers in the desert." – Isaiah 43:19

I knew from less than a week of texting her that she was a free-spirit, open-minded Christian like me. What I didn't know, was that the card I happened to pick out…along with the verse that came to mind to write in it on the spur of the moment…was perfect. Christine had a tattoo on her ankle of a lotus flower. She knew what Namaste meant…and it was important to her. She also knew the Bible verse and had been holding onto it as a promise. We were both divorced Christians hoping for a new beginning.

Spiritual and secular culture seem to be coming together, and there is an increasing use of the word *namaste* (pronounced \NAH-muh-stay). The term is associated with both Hinduism and yoga. The word comes from Sanskrit and literally means "I bow to you", and is used as a greeting. Sanskrit is the ancient and classical literary language of Hinduism. Other well-known borrowings from Sanskrit in English include the words karma and nirvana.

I looked up Namaste online and found my favorite version for its meaning on a *Pinterest* post as follows:

"My soul honors your soul. I honor the place in you where the entire universe resides. I honor the light, love, truth, beauty & peace within you, because it is also within me. In sharing these things we are united, we are the same, we are one."

This definition of Namaste, I think, coincides well with the "fruits of the Spirit" listed in *Galatians*, and has a close connotation to the true meaning of the word in *Genesis* that got translated "to have dominion over". Another word that conveys a similar principle is the Hebrew word for peace: *shalom*; which is derived from a root denoting wholeness or completeness. Its frame of reference throughout Jewish literature is bound up with the notion of *shelemut*: perfection. Its significance is thus not limited to the political domain — to the absence of war and enmity — or to the social — to the absence of quarrel and strife. It ranges over several spheres and can refer in different contexts to bounteous physical conditions, to a moral value, and, ultimately, to a cosmic principle and divine attribute. (https://www.myjewishlearning.com/article/shalom/)

Words and symbols mean different things to different people. But, most people probably mean something good by them…maybe even the same thing.

Displaying cool words on bumper stickers and shirts is easy. Living out the meaning of the special words and symbols we embrace is more difficult. There are myriad ways to divide people into opposing camps and to "stir up people" into conflict…but only one way to unite them…LOVE.

When I was a young forestry student, one particular student would constantly ask questions in class that made the rest of us gasp and smirk. He outdid himself one day when he stood up in class…shaking with nervousness and anger…and asked: "What about the spirits of the trees!? When considering how to manage the forest, do we ever account for the spirits of the trees?!"

I don't know what happened to that student. I don't know what he meant, and I couldn't see any sense in considering the "spirits of the trees". How would you do that anyway? But, his question did make me think about how I often felt when I stood in a mature forest; especially when with another like-minded person. My mind immediately portrayed in front of my mind's eye the time my Grandpa described to me how he felt he was in church under the canopy of an undisturbed mature forest, and how I took comfort in knowing he felt it too. Then, the image of this place destroyed by the timber thief, was as a desecrated temple to my inner man. Yes, standing in a mature forest gave me the same feel as if standing in a great Gothic cathedral with its arching design and stained-glass windows. I felt it, too, in the Roofless Church at New Harmony and in the little abandoned church in rural Jasper County. It isn't the "spirit of the trees", it was The Spirit that we need to consider when managing forests; as we are connected, in spirit, with not only the trees, but with all of Creation.

There is a book called *The Spirits of the Trees: Science, Symbiosis and Inspiration* (2006) written by harpist and artist, Fred Hageneder. The book explores the relationship of humanity and trees and their ability to sustain, sooth and inspire mankind. Fred is the founding member of the German/French charity *Friends of the Trees*, which helps with the creation of modern sacred groves… places largely cut down as pagan Europe was Christianized…as spaces of tranquility as well as multicultural and multi-faith meeting places. Maybe considering the "spirit of the trees" is not so crazy after all. Maybe it has as much to do with considering our own spirit as it does anything else. Maybe our own spirit feels this sense of sacredness in a grove of high canopy trees because we sense The Spirit in such places.

Jesus often went into the wilderness, and prayed in the Garden of Gethsemane. Yet, how many churches provide such a place for their members or the community at large? The Abbey of Our Lady of Gethsemani is a monastery near Bardstown, Kentucky started by French and Swiss monks. It is a part of the Order of Cistercians of the Strict Observance (*Ordo Cisterciensis Strictioris Observantiae*), better known as the Trappists. Gethsemani was founded in 1848 and raised to an abbey in 1851 and is the oldest monastery in the United States that is still operating.

Thomas Merton, bestselling author of the book *The Seven Storey Mountain* (1948) was a monk at Gethsemani. Throughout his life there, he studied Buddhism, Confucianism, Taoism, Hinduism, Sikhism, Jainism and Sufism in addition to his monastic studies. He lamented how Christianity had forsaken its mystical tradition, and observed that the human experience of "being" was still largely intact in the eastern philosophies; which were mostly untainted by Cartesian thinking. He had studied the writings of the Desert Fathers and other Christian mystics and found many parallels between their writings and Zen philosophy. Merton's writings still inspire people today. Last year, there was a program offered at Bellarmine University titled: "Awakening the Creative Spirit: A Day Inspired by Thomas Merton".

Gethsemani is one of the field trips offered with the Society of American Foresters meeting in Louisville, KY this fall (2019). Their 2,000 acre property has walking trails which feature meditative places with scriptures and statues in the forest; while the grounds around the monastery are beautifully landscaped. If there were any foresters in Kentucky before Silas Mason, I suspect they were at Gethsemani…humbling going about managing the forest there with no fanfare.

I have observed more examples of beautiful landscaping and reverence shown for nature in Catholic churches than any others. The rural Lutheran churches I grew up around had groves of trees where they would hold church picnics, but most of these have went by the wayside in favor of fellowship halls now, as folks expect to be where there is air conditioning now. I suppose the impact of St. Francis among Catholics is still evident to some degree.   St. Francis reportedly believed that people have a duty to protect and enjoy nature as both the stewards of God's creation and as fellow creatures ourselves. The One-ness of Creation was understood by him and his example has lived on to some degree in some places like Gethsemani. Where the creative spirit is awake, there is always both beauty and utility…art and science…in harmony.

Why don't we have more "Gardens of Gethsemane"? Few churches pay attention to their landscape and grounds in a way that might inspire stewardship and a creative spirit and provide a place for it to flourish. My wife and I are making it our mission to try to change that. We hope to help facilitate a "natural bridge" to try to span the gap that seems to exist between church folk on one hand and environmentally aware folk on the other.

The idea that One-ness within Mankind and Nature is "One-derful" and timely. Trees have been shown to have "social networks", and that the idea that there could be some semblance of "spirit" attributed to trees, has been presented in popular books over the last few years. People are beginning to want to understand these concepts and unite in meaningful ways to engage with one another to help repair the damage Mankind has caused to the planet. People are looking for ways to heal the damage caused to its people as well. What if the social networks of trees are communicating with us? What if we belong as a part of this community too…and this sense of belonging is being communicated to us through these awe-inspiring sensations so many of us feel "in our spirit"? What if this is the way we were made…the way we are supposed to be?

"For ye shall go out with joy, and be led forth with peace: the mountains and the hills shall break forth before you into singing, and all the trees of the field shall clap their hands." - Isaiah 55:12 (KJV)

The art of forestry is the heart of forestry. To me, it is akin to tending a garden…an endeavor that harkens back to tending the "Garden of Eden". The Garden of Eden, like the Garden of Gethsemane, had trees. Gardening, farming and forestry all go together and they are, I would argue, what we were created to do. When we are active in nature; participating with its cyclical rhythms and interacting with other life forms, we immerse ourselves within Creation and find ourselves at home. We hear the birds sing, and her the subtle songs the breeze plays as it blows through the trees. Who can deny that there is a strong connection between a healthy land and healthy human relationships? Neurologist and author Oliver Sacks writes in his short essay titled *Why We Need Gardens*, found in *Everything in Its Place: First Loves and Last Tales*:

"As a writer, I find gardens essential to the creative process; as a physician, I take my patients to gardens whenever possible. All of us have had the experience of wandering through a lush garden or a timeless desert, walking by a river or an ocean, or climbing a mountain and finding ourselves simultaneously calmed and reinvigorated, engaged in mind, refreshed in body and spirit. The importance of these physiological states on individual and community health is fundamental and wide-ranging. In forty years of medical practice, I have found only two types of non-pharmaceutical "therapy" to be vitally important for patients with chronic neurological diseases: music and gardens."

Neuroscience is making some discoveries that are able to quantify how connected we really are to our environment. Dr. Candace Pert has demonstrated that there is a biomolecular basis for our emotions. Her book Molecules of Emotion, explains how molecules carry literal photocopies of thought formulated in the memory networks of the brain. Information molecules are actually able to cause changes at the cellular level…even to our DNA…and cause disease. Negative experiences such as physical or emotional abuse release negative chemicals while positive experiences release positive chemicals… building an unhealthy or healthy network. These memory networks of the brain are comprised of "forests" of tree-like nerve cells which have been named "magic trees of the mind" by another groundbreaking brain researcher, Dr. Marion Diamond. (Leaf, 2007).

Dr. Caroline Leaf describes how we have around 100 trillion of these magic trees in our brain and each one is capable of growing up to 70,000 dendrite branches where memories are stored. When we forget something, it is because the branch that contains that memory has been literally pruned off by a glial cell. Glial cells are the "foresters" working among our "magic forest of the mind". They provide nourishment and protection to the growing "magic trees" and they cleanse the forest of waste material and prune branches containing unnecessary or unwanted memory. Dr. Leaf's work is dedicated to helping people to detoxify their brain through developing a controlled thought life; which includes learning to balance your work and rest, deep breathing, exercise, massage, meditation, prayer, sleep, visualization, prayer and practicing forgiveness (Leaf, 2007). We can train our glial cells…the foresters managing the forest in our mind…to prune away the harmful branches.

Beautiful settings in nature provide the peaceful, inspiring places in which we can best develop healthy thoughts. Maintaining these areas, we get exercise. Working together, we can develop healthy relationships built upon a common goal. Doing this, we achieve community and we build together a "land ethic" along the way. I'd love to see a nationwide effort on the level of the Civilian Conservation Corps, to do this on a large scale. My grandpa once said he "discovered what it was to be an American" after this experience. The CCC's trained so many people from so many backgrounds with skillsets and a sense of camaraderie that was essential to them and their country in the decades to come. It helped build the "greatest generation".

Meanwhile, in the absence of such an effort, we have to do this ourselves… on whatever scale we can, in our own communities. Imagine if every little town, every neighborhood of large cities, had a community garden, orchard, forest, and an inspiring meeting place in a similar spirit of the Roofless Church would be collaboratively and creatively built with purpose. Together, the marriage of science and the arts create the spiritual offspring needed for us to not only survive, but to thrive…in body, soul and spirit…in harmony together and with the land. Agricultural, silvicultural, artistic, and spiritual pursuits could all occur in such places spontaneously, organically and without the danger of one powerful person or political group subverting it. According to Dietrich Bonhoeffer; who became a martyr helping save Jews from the holocaust:

"The person who's in love with their vision of community will destroy community. But the person who loves the people around them will create community everywhere they go."

———————

An enduring land ethic will necessarily inhabit all three parts of the person: mind, body and Spirit. We learn. We do. We feel. When all three are in tune with each other, we know how to love, to heal, and help others. In our hearts, we carry this capacity to heal the land and each other. We also carry the capacity to abuse the land and each other. As Solzhenitsyn concluded:

"…it's not about one system over another…one philosophy or nationality or

religion over another…the line dividing good and evil cuts through the heart of every human being. And who is willing to destroy a piece of his own heart?"

The book, *Bury My Heart at Wounded Knee: An Indian History of the American West*, by Dee Brown, came out in 1970…the same year I was born. It was a time of increasing American Indian activism. The title is taken from the final phrase of a twentieth-century poem by Stephen Vincent Benet titled American Names; which reads:

"I shall not be there. I shall rise and pass. Bury my heart at Wounded Knee."

Benet's poem is not about the plight of Native Americans, but was fitting because of the Wounded Knee Massacre…where some 250 Lakota were killed… mostly women and children. It was the last major confrontation between the United States Army and Native Americans. Incredibly, at least twenty of the soldiers were awarded the Medal of Honor.

Yes, it is difficult for a person to "destroy a piece of his own heart", but it is difficult to imagine those twenty soldiers accepting a medal of honor for firing on women and children without having first destroyed a piece of their own heart: the good piece. Unfortunately, there are plenty of instances of this throughout history.

To destroy, instead, the piece of one's heart where evil resides; this should be the aim of us all. Then, whatever we have done; whatever our culture or our nation has done; we can answer for it with: "I shall not be there. I shall rise and pass. Bury my heart at (fill in the blank)". The heart thus planted becomes as a seed to be "born again" to grow into something beautiful and glorious. As a Christian, I believe this is what we are called to do, spiritually, when we are "buried in baptism"…a symbolic burial of the "old man" and resurrection in the spirit as a "new man" in Christ. *(Romans 6:4)*

I have written a lot about science, art, and spiritual matters. I've portrayed "know-how" and "elbow grease" as the core essentials in getting things done. Yet, at the end of the day, we have to have a place to do stuff. Some land is needed on which to practice forestry, gardening, farming, and crafts. Land is needed to operate a working community-aligned forest, garden, and/or farm. If we are to establish a land ethic, we must have access to land!

In the Nordic countries where there is a long-established "right to roam", their cultures developed along the premise that the forests, lakes, rivers, shorelines, and so on are to be experienced and enjoyed…and treated with respect… by everyone. Germany's "right to roam" law is called *Betretungsrecht*, and is guaranteed by federal law. The Constitution of Bavaria, where I was on my forestry trip, guarantees everyone: "…the enjoyment of natural beauty and recreation in the outdoors, in particular the access to forests and mountain meadows, the use of waterways and lakes and the appropriation of wild fruits". Nick-named *Schwammerlparagraph* (mushroom clause), it obliges everyone to treat nature and the landscape with care. Doesn't that sound wonderful!?

I look at the "right to roam" as a survivor of the indigenous mindset. It survived in other places too…but mainly in barren lands at the fringes of empires where the inhabitants were driven into survival mode. For instance, the Scots Irish who populated Appalachia along with Germans, came with nothing. Oppressed for centuries and driven off their fertile lands into the barren wastes, they were then driven even from Scotland and planted in the north of Ireland against the will of the Irish who were already there, by the British Empire as serfs. Coming to America, a similar situation was encountered. The Germans, on the other hand, largely immigrated as intact communities of skilled trades-men with sufficient funds to buy land or soon acquire it through cooperation together as a community. I have ancestors from both; as well as Native Ameri-can.

It seems to me that Germany became the place in the West where forestry developed because its people survived empires long enough to thrive into the industrial age and become a powerful nation. But, at the same time, they were thwarted in becoming an empire. People must have sufficient resources and a secure outlook for the future to manage for the long-term. The destitute can't do that, and the powerful don't have to. Empires and the empire economy have stripped the indigenous culture based on cooperation, sharing and creating… and replaced it with a system based on competition, stealing, and dismantling. Reductionist science has become their tool; reducing everything into material resources to be analyzed in minutia to obtain the greatest profit with no consid-eration for the whole.

The empire economy extends its tentacles into every corner of the planet to find somewhere it can get resources; even "human resources". Supply chains connect global networks of chain stores and companies to cheaper resources, lower regulatory environments and more defenseless populations. These chains can enslave populations as surely as iron chains. The goal is to always get something easier, cheaper, or at someone else's expense. Loving each other is replaced with using each other for the love of money…the "root of all kinds of evil". Empires have pushed their people into doing this until they don't know how to do anything else.

All peoples of the earth are of one blood, but not of one mind. To dig up all the ginseng before someone else does, to cut all the timber while you can get it, to litter because others do, to look only to yourself and not to the wider com-munity, becomes a normalized survival behavior by the empire model. The Bible calls this system *Babylon*. It leads to apathy, anger, suspicion, discontentment and depression. Loneliness ensues, and then emptiness. Yet, the mind can be renewed.

We sense it. We long for it. When we stand under a cathedral-like canopy of mature forest and hear the birds chirping above us in the gently swaying branch-es…when we look out upon the sea and hear its crashing tides and smell its salty spray…when we peer up at the vast multitude of stars on a clear, chilly night in the desert…we know it exists. This is our heart telling us to listen, to look to it for what has been drained from our mind; forgotten. Our heart has access outside ourselves. If we listen to our heart and give it permission to fill our mind, we can begin to learn afresh this indigenous mindset. With all the accumulated knowledge science has given us over millennia, we better use our hearts to manage it for the common good, for a new beginning.

We don't have the "right to roam" in America…and I don't think we'll get that anytime soon. I wouldn't even advocate that here, as it would impose upon landowners to allow the public to trespass on their land. A land ethic has to become commonplace to harmoniously accommodate a wholesale "right to roam" on private lands. But, we do have National Forests, National Parks, wildlife refuges and city parks. America is a leader in having established these wonderful places and we're blessed to have these treasures. We have more wilderness than most other countries. These are to be cherished and enjoyed… and fought for as continued public resources. Strict Constitutionalists will say that the Founding Fathers did not intend for the Federal Government to own such vast lands, and they are correct. But, to say that this means we should sell them off is to ignore that the government was also intended to be ran by "We the People". That's us!

Our nation's public lands are great places to roam and they are great places to volunteer time as well. These places are a great place for a "land ethic" to be practiced and developed too. More of our youth need to have access to these places and programs to help them experience hiking, camping and nature exploration.

Public lands, however, do not provide the same function as working com-munity forests or farms. They cannot engender a "land ethic" like community forests because they are not community-based, place-based, working lands. We need to establish places for that to happen. Wilderness areas are where we should learn to "leave no trace", while community forests are where we can learn to "leave a positive trace". Here, we can see people positively impacting the environment in various ways. Each place would have its own activities, oppor-tunities, and examples, as every locality is different.

The "right to roam" need not become immediately applicable everywhere… but it does need to become available to everyone somewhere. When folks have the right to roam physically, I believe, it will help engage our minds to roam as well. Confined, we stagnate, but set free we can learn to soar. No matter where you live, there should be a place where you can get your hands dirty in the forest, in the garden, on the farm. This is how a "land ethic" can develop across the land and make it possible so that perhaps one day a "right to roam" will be welcomed on a wide scale by an ever-widening scope of landowners. It should not be in any way forced.

To create a community forest, or even a community garden, it takes land and resources. This requires money and/or the generous giving of land. Berea College owes its existence to philanthropy. Since Cassius Clay donated ten acres to John G. Fee in 1853, many generous folks have donated large and small contributions to insure that Berea College prosper and continue to provide a tuition-free education to students who most likely would not have been able to attend college otherwise. The Berea College Forest, similarly, owes its existence to philanthropy. Owned by a free-tuition College, the Berea College Forest is free to visit and enjoy by all. This is only possible through generous giving.

Silas Mason, the first Berea College Forester, made the first land purchase in 1898 with his own funds. A pioneer in American forestry, Mason understood that forests could be, and should be, sustainably managed for future generations and that education was the key to conservation. Over the next twenty years, due to the generosity of Ms. Sarah Fay, of Boston, the forest holdings had reached 5,400 acres. She also provided financial support to the Appalachian Mountain Club…which gave rise the Appalachian Trail…and donated land in New Hampshire; which became the Fay State Forest. Ms. Fay's father, Joseph Story Fay, was a businessman and philanthropist who gave land and funds to build a church and schools, and was a renowned lover of trees. He reforested much of the landscape around Woods Hole, Massachusetts, where he owned a summer residence and land. We need people like Silas Mason, but we also need people like Sarah Fay…and they need to be able to find each other.

The Bernheim Forest, near Louisville, is another well-known Kentucky forest that is enjoyed by many. Established in 1929 on land donated by German whiskey-distiller Isaac W. Bernheim to the people of Kentucky for a gift. It is clear to me that Mr. Bernheim knew the value of the "right to roam" and hoped to create a place for folks to enjoy this in his new homeland:

"…I have expressed my intention that said property… be held in trust… and said fourteen thousand acres be used for a park, for an arboretum, and, under certain conditions, for a museum, all of which are to be developed and forester maintained… for the people of Kentucky, and their friends, as a place to further their love of the beautiful in nature and in art, and in kindred cultural subjects, and for educational purposes, and as a means of strengthening their love and devotion to their state and country." (Bernheim Forest website)

I find the quote by Bernheim fascinating and complete. He twice mentions love: love of the beautiful in nature and in art… and love and devotion to state and country. There is wisdom in his statement. I hope that the Berea College Forest, as well as the Bernheim Forest, does provide a place in which people can further their love of these things. Fortunately, philanthropists have cared over the years enough to provide some of our most treasured places enjoyed today.

Leo Drey, of St. Louis, Missiouri, is another notable philanthropist. Drey was the largest private landowner in Missouri until he gave away all of his property for the enjoyment of others. In 2004, Drey donated his 146,000 acres of Ozark forest lands…called the Pioneer Forest…to a charity that will continue his mission of sustainable forestry. Recently, Mr. Drey passed away and he even donated his body to the Washington University School of Medicine. The Pioneer Forest continues to be a leader in creative forest management and a testament to the viability of unevenaged management.

Another notable St. Louis philanthropist is Henry Shaw. He donated land for Tower Grove Park in St. Louis, endowed Washington University's School of Botany, and helped found the Missouri Historical Society. He gave the city a school and land on which to build a hospital. Shaw is most famous for establishing one of the world's first and most highly regarded botanical gardens. Working with leading botanists of his time, he planned, funded and built the Missouri Botanical Gardens in 1859. Today, the Garden is a National Historic Landmark and a center for conservation and education. (Missouri botanical garden website).

We hear a lot about the "robber barons" of the Industrial Age, and we certainly hear plenty about corporate greed today. However, it is refreshing to know that many individuals with great wealth have been very generous too. Conservation and education have a lot to thank for philanthropy. Philanthropy can come in any size, too. Berea College continues to rely on donors large and small to provide a tuition-free education to our students and to fund many of our initiatives. I'd be happy to see our efforts continue to grow this way here. But, I am thinking on a larger scale than that in writing this book.

What if every county had a community forest? It was the dream of College President Frost…noted on the last page of "Our Southern Highlanders" (1913)… for a "model farm" to be established *in every mountain county showing how to get the most out of mountain land"*. Mountain land across Appalachia is largely forested, and such model farms would necessarily also be model working forests. There is so much absentee forest land across Appalachia that would be great for this. Such community farm-forests could be established, of any size, similar to the Berea College Forest. These forest-farms could provide multiple benefits and sources of wood for their local economy. I would envision some of these places having draft horses and portable sawmills as a part of their operation. A myriad of other creative endeavors, no doubt, would arise through place-based application of creative ideas. I hope there are some modern day Silas Masons and Sarah Fays out there today who need only to receive a vision to make such places become a reality in more communities. Even a vacant lot in town could serve as a valuable resource in some communities and provide a catalyst for action and inspiration where we can "be the change we want to see in the world".

For a land ethic to take root in Appalachia and beyond, the people must have access to the land. I'm not talking about recreational areas, as in parks. Those are great and they fill a purpose; but community farms and forests where people can learn how to work the land is what is needed so that agri-culture and silvi-culture and a local economy can thrive again. The land, the plants and trees and fish and animals, all of us and our human culture can thrive, must thrive, together. We can learn to create something, live again, love again, heal ourselves and each other…and heal the land in the process. This will take people of all skillsets and backgrounds, people with vision, with heart, and with the means to make it happen.

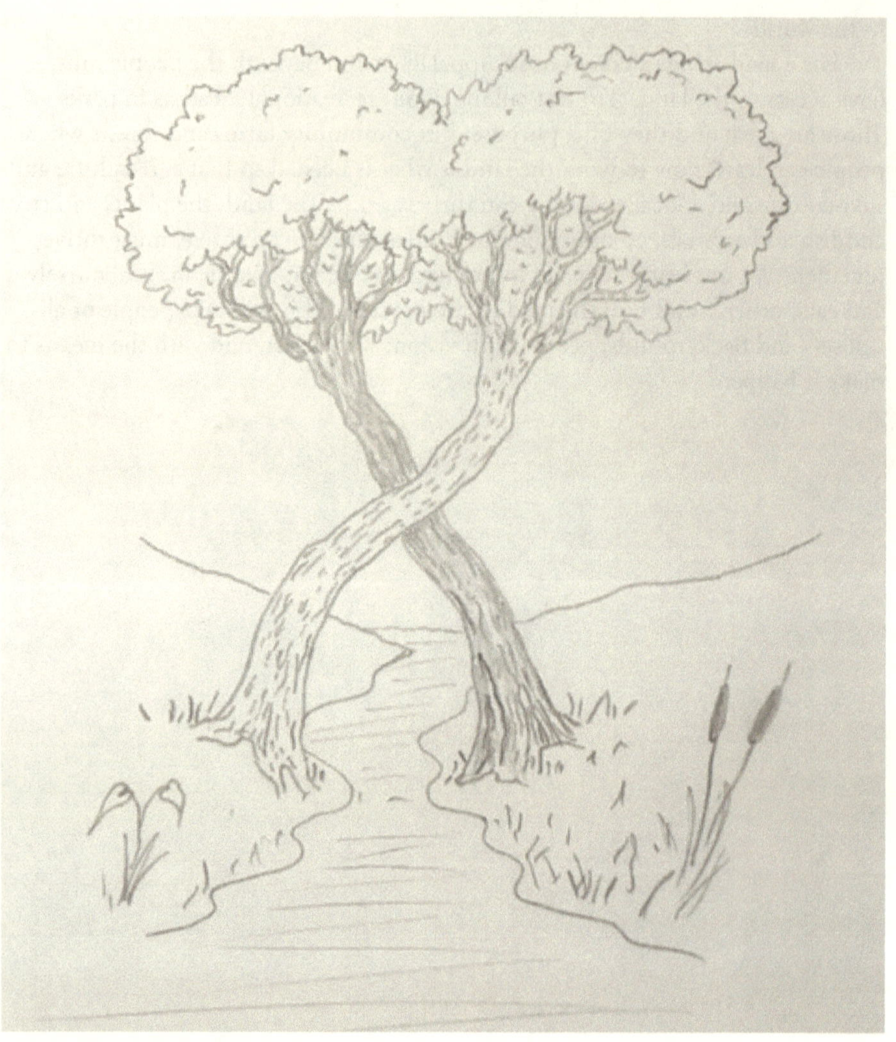

Parting Thoughts

Music and gardens…the arts and nature…are vitally important to our wellbeing. Healthy, active, whole people are needed as stewards to help revitalize our environment and communities. This is the circle of life we find ourselves in. A certain theme song for this book comes to mind for us to gather to: Iron Butterfly's 1968 song *In-A-Gadda-Da-Vida*.

This song is sometimes touted as the first "heavy metal" hit. The singer was actually saying "In the Garden of Eden" in such a garbled voice that it came out sounding like "In-A-Gadda-Da-Vida" and they kept it that way. Here are the main lines to the song, with the chorus changed to "In the Garden of Eden":

"In the Garden of Eden, honey; don't you know that I'm loving you? In the Garden of Eden, baby; don't you know that I'll always be true? Oh, won't you come with me; And a-take my hand? Oh, won't you come with me; and a-walk this land? Please take my hand."

The song invokes an image of Adam and Eve walking, in wonder, in the Garden of Eden when all was well. On another level, I imagine Mankind (men and women) asking Nature for another chance to be the husbandman of the Earth…as equals, as One Body, the "Bride of Christ" this time. This is the feeling my wife and I get when we walk in Nature together. Love and Knowledge combined produces the Wisdom to manage this Earth as we were created to do…restoring Order as the Highest Virtue…according to *Gottesblick*… until *"Thy Kingdom come, Thy Will be done, on Earth as it is in Heaven."*

The day after Christine and I went on a hike at Rocky Fork for our first day, she took me to her favorite childhood place in Nature…Bays Mountain. The wolves put on a display in front of us. It felt like they were showing us their "green fire" as Aldo Leopold described seeing in the eyes of a dying wolf. They even howled as we left. When I took her to my favorite childhood place, "The Crick", we walked until we reached the extent of where I commonly went as a child…the last bend where a car horn could still be heard. After forty years of changing sand formations, I was surprised to see the sand bar was back to the same shape as I remembered it when I was collecting rocks here and hastily left a stopped up bottle. Looking down, I saw the bottom of a thick, old glass bottle barely sticking out of the sand. Picking it up, a smile came to my face as I remembered my nine year old thumb struggling to push a rock into it. As I washed it off, I knew in my heart it was the same bottle. There it was again… unbroken and with the rock stopper still in it.

Glass bottles used to be turned in…redeemed…and filled back up. We may find ourselves feeling discarded and empty…divorced from our former spouse, from our family, from our community, from Nature, from God. But, a new beginning is possible. Our bottle, our spirit, our culture, our environment, can still be redeemed and filled with New Life. A new beginning where we work together, in love, to participate in the healing and repair of Nature, as well as of Humanity is in front of us. Steinbeck said it well, on the Sea of Cortez: "None of it is important or all of it is."

So, *"Oh, won't you come with me; And a-take my hand? Oh, won't you come with me; and a-walk this land?"*

If you would like to see what we are up to, check out our website: www.pneumanoria.com

Namaste.

Bibliography:

Alexandersson, Olaf (2002). Living Water: Viktor Schauberger and the Secrets of Natural Energy. Gill Books.d

Becker, Rachel (2015). This Bonsai Survived Hiroshima, But Its Story Was Nearly Lost. National Geographic. https://news.nationalgeographic.com/2015/08/1508050-japanese-bonsai-survived-hiroshima-bombing/

Bohm, David (1980). Wholeness and the Implicate Order. Routledge.

Bordewich, F. 2005. The ambush that changed history. Smithsonian magazine, Sept.

Braun, Werner von (1952). Das Marsprojekt. Frankfurt: Umschau Verlag.

Cotta, Heinrich (1902). "Cotta's Preface", Forestry Quarterly, vol. 1, no. 1, p. 5. (tr.from *Anweisung zum Waldbau*, 1817.)

Cohen, Adam (2017). Imbeciles: The Supreme Court, American Eugenics, and the Sterilization of Carrie Buck.

Coyle, Daniel (2009). The Talent Code, Bantam Dell, New York, NY.

Darwin, Charles (1859). "On the Origin of Species by Means of Natural Selection, or the Preservation of Favoured Races in the Struggle for Life," p. 155.

Denton, Michael (1986). "Evolution: A Theory in Crisis," p. 250. Adler & Adler.

Eakin, Hugh (November 2000). Picasso's Party Line. ARTnews. Vol. 99 no. 10. Archived from the original on 25 July 2011.

Hageneder, Fred (2006). The Spirit of Trees: Science, Symbiosis and Inspiration.in

Hamilton, Edith (1938). The Greek Way. W.W. Norton & Company, Inc., Time Reading Program Special Edition, Time Incorporated, New York.

Jackson, Wes (2000). The Changing Relationship Between The Tree of Knowledge and the Tree of Life, The Land Report, Issue No. 68, Fall 2000.

Leaf, Caroline (2007). Who Switched Off My Brain? Controlling Toxic Thoughts and Emotions. Dr. Leaf.

Leopold, Aldo (1949). Sand County Almanac. Ballantine Books. New York.
Newton article: (nationaltrust.org.uk)

Patterson, Clint (2017). A Century of Forestry. Berea College.

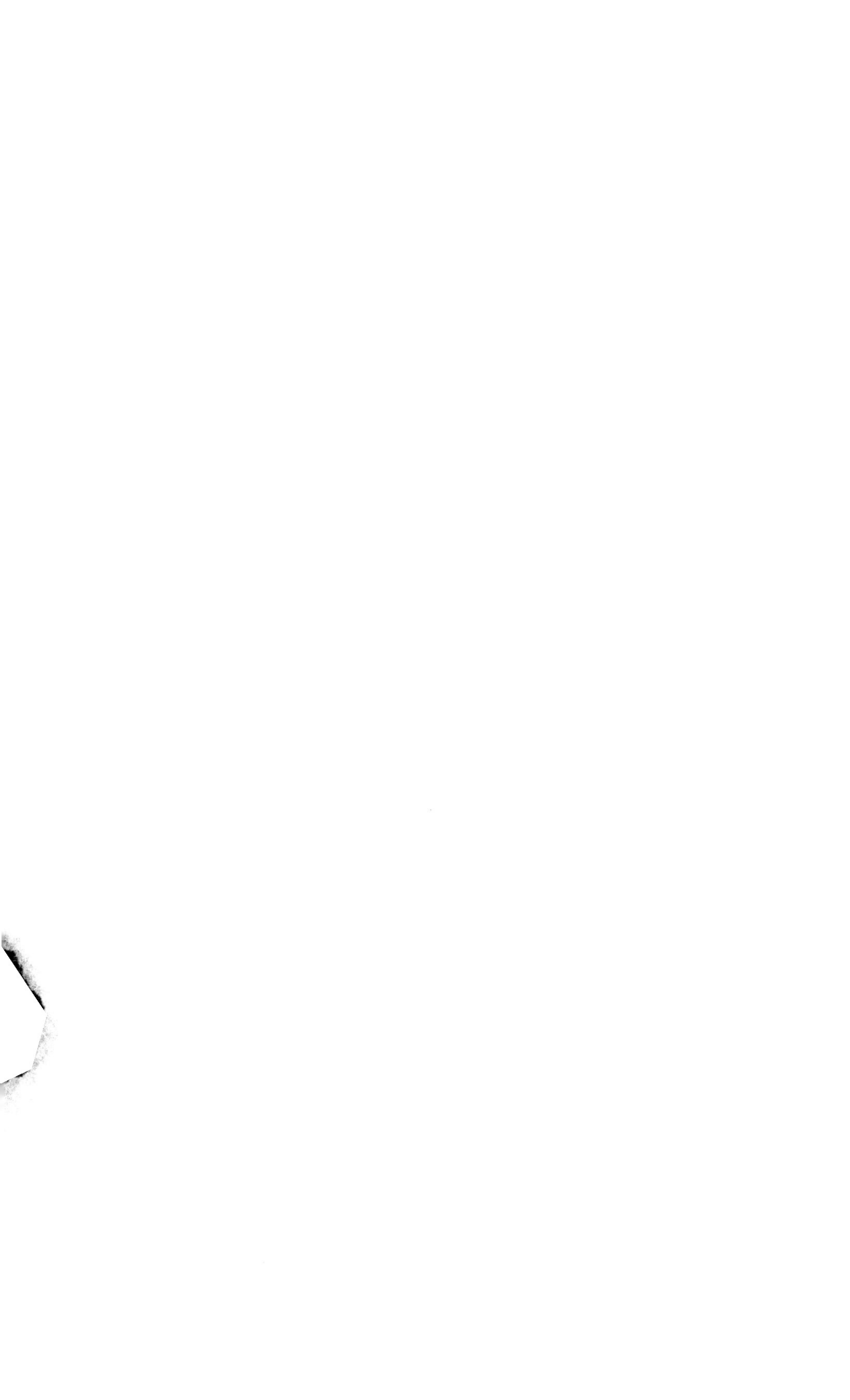